T0074742
Pecten arcuatus
Avicula echinata
Avicula costata
Isocardia minima
Cardium cognatum
Trigonia impressa
Lima proboscidea
A. inæquivalvis
Trigonia striata
Geromya concentrica
Modiola gibbosa
Astarte elegans
Modiola cuneata
Modiola plicata
Gervillia lanceolata
Nautilus truncatus
Ammonites serpentinus
Amm. bifrons
Belemnites tubularis
A. communis

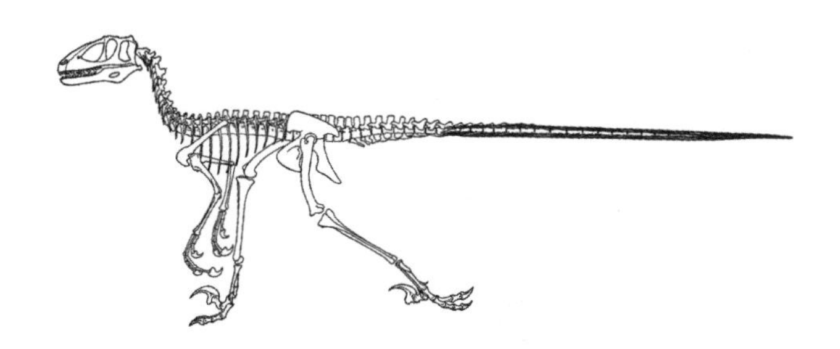

# PALEONTOLOGY

## AN ILLUSTRATED HISTORY

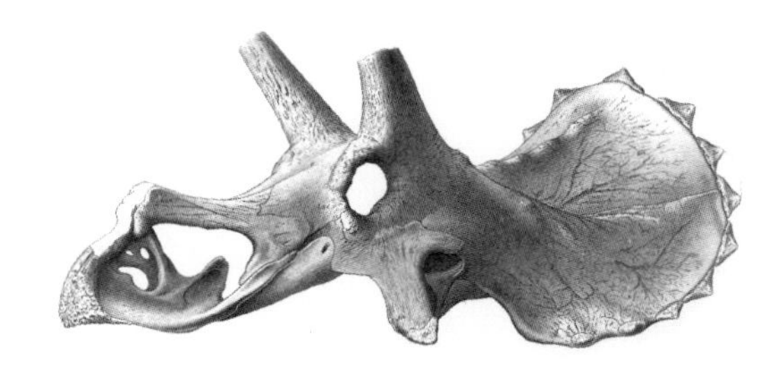

# PALEONTOLOGY

## AN ILLUSTRATED HISTORY

DAVID BAINBRIDGE

PRINCETON UNIVERSITY PRESS
PRINCETON AND OXFORD

Published in 2022 by Princeton University Press
41 William Street, Princeton, New Jersey 08540
6 Oxford Street, Woodstock, Oxfordshire OX20 1TR
press.princeton.edu



Library of Congress Control Number: 2021941945

ISBN: 978-0-691-22092-5
Ebook ISBN: 978-0-691-23592-9

Conceived, edited, and designed by Quarto Publishing plc,
an imprint of The Quarto Group

Typeset in Adobe Caslon and Gotham

Design by Blok Graphic, London
Editor: Caroline West
Projects editor: Anna Galkina
Deputy art director: Martina Calvio
Picture researcher: Sara Ayad

Printed in Singapore
10 9 8 7 6 5 4 3 2 1

---

**Page 1: *Deinonychus* skeleton, illustration by Robert Bakker, from *Bulletin of the Peabody Museum of Natural History*, Yale University, 1969, no. 30, p.142.**

---

**Page 2: *Hypsilophodon*, illustration by Neave Parker, 1960. Natural History Museum, London.**

---

**Page 3: John Bell Hatcher (1861–1904), *The Ceratopsia*, 1907; Skull of *Triceratops serratus*.**

---

# CONTENTS

# FOREWORD: THE PLEASURE OF RUINS

Writing this history of paleontology has revived vivid fossil-related memories from my past. Ever since childhood I have been fascinated and inspired by the fragmentary vestiges of ancient creatures, the rubble of once great nations of animals.

Every birthday I would plead with my parents to be taken to the Natural History Museum in London, in a more naive era when a small boy could be set free to explore the terracotta halls of this Romanesque "cathedral of life." I loved the grand dinosaur skeletons, of course, although I was already aware that some of them were made of plaster, but it was the unfrequented gallery of ancient fish that I remember most. Even more than their spectacular land-living descendants, this gallery's inhabitants looked intriguingly, *overwhelmingly* old—crushed and mangled messengers from an alien era before feet trod the Earth. Many appeared timid and harmless, yet some were heavily armored, hinting that some awful threat was abroad in those days.

A decade or so later, I started to train as a veterinarian and soon the structure, the architecture, of animals began to grip my attention. Every

**Heracles, Hesione, and the Trojan Cetus (or the "Monster of Troy"); Corinthian ceramic black-figure krater, circa 550 BCE. Museum of Fine Arts, Boston.**

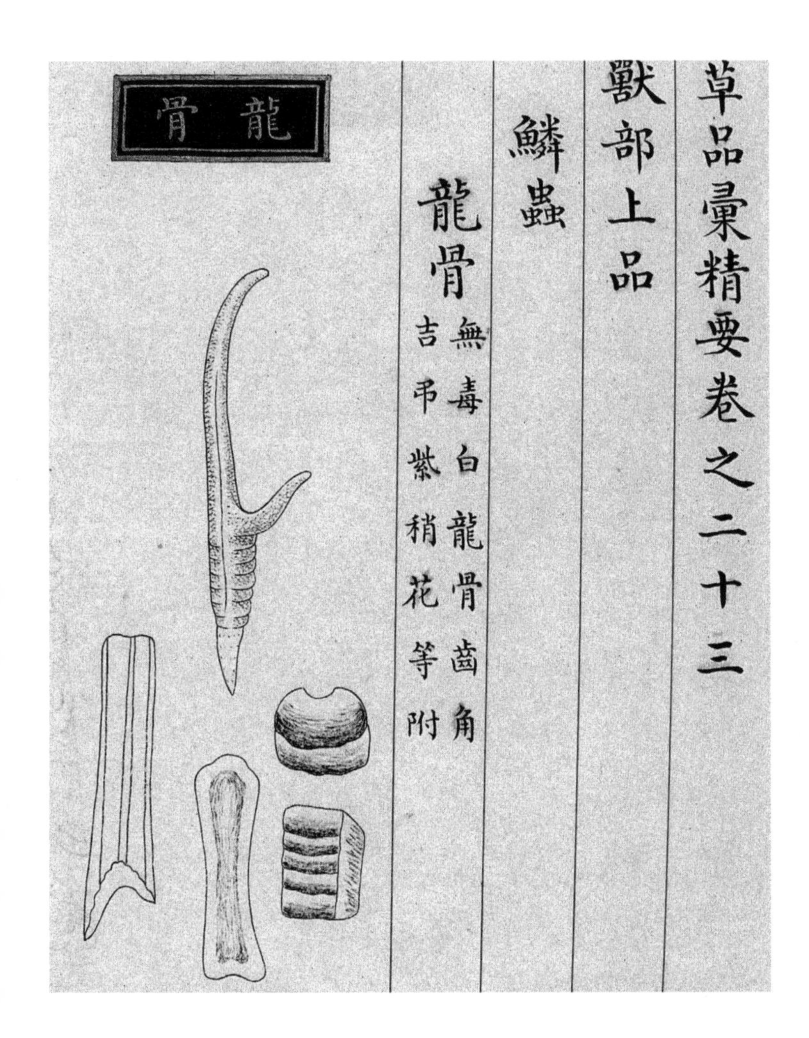


**Liu Wentai and others, *Bencao pinhui jingyao (Materia Medica)*, 1505; Dragon bones.**

anatomical strut, hinge, and tuberosity appeared wonderfully adapted to fit its function, the living embodiment of the physics of making a flesh-machine work. I knew animals had never been "built" as such, but the unthinking meanderings of natural selection had clearly yielded constructions an engineer would understand. While at university, I was able to study a discipline of my choice for a whole year, and this is why I ended up with a degree in vertebrate and mammalian paleontology—and it just so happened to be a year when many new fossils and many new ways of thinking about evolution were afoot. I was also able to study for a summer at Cornell University, in upstate New York, and ensured that I fitted in a trip to the American Museum of Natural History in New York City, where I was delighted to find the dinosaur skulls are not made of plaster, and are so numerous they can be mounted communally on the walls in giant family-tree tableaux.

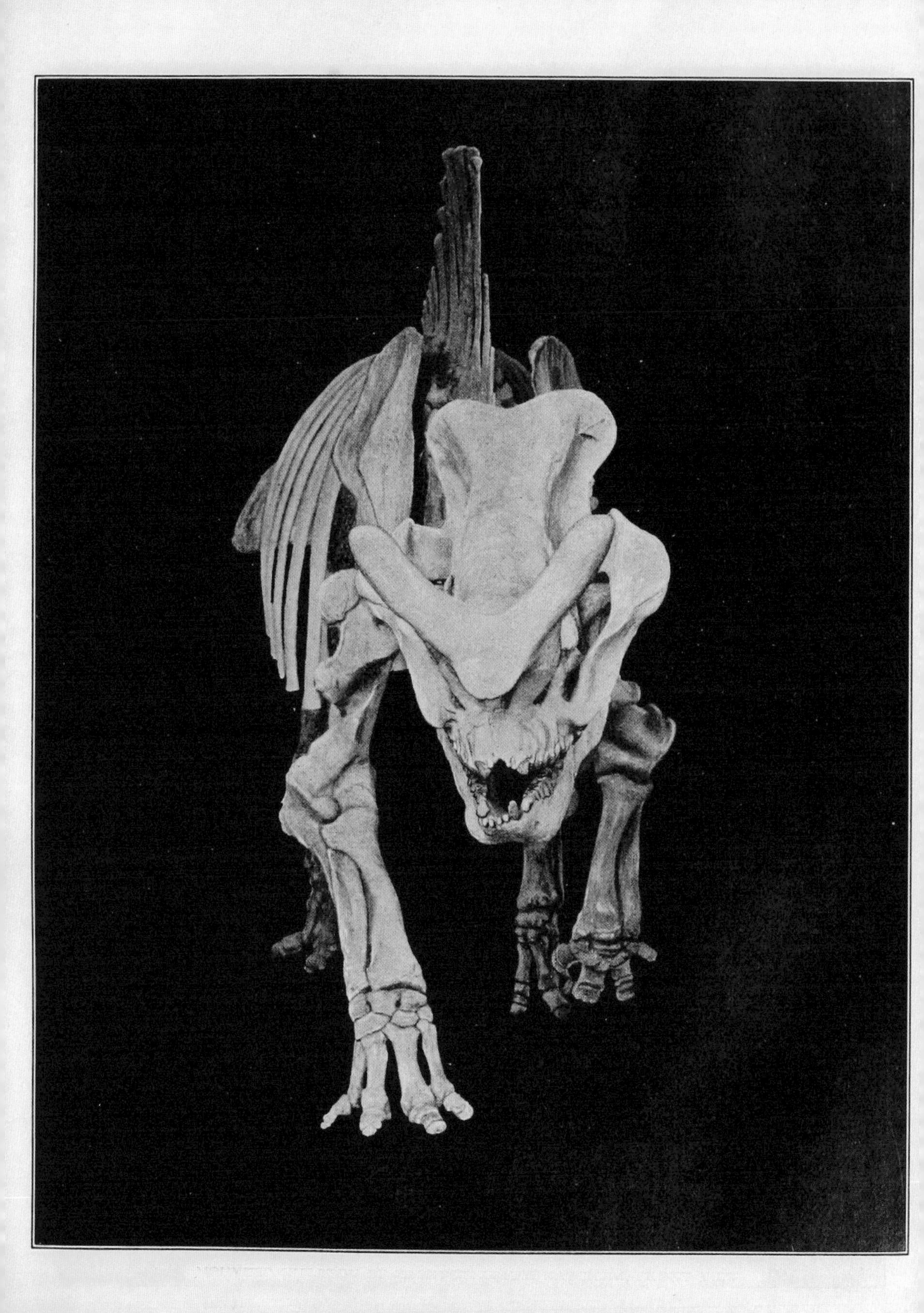

So now, thirty years later, my aim in this book is to tell the history of the discipline of paleontology, ironically a relatively young science—perhaps only two hundred years old in its fully scientific form. On the way, I hope to show how fossils are visually striking in their antique, disarrayed way, and that they have inspired some striking images, too. I will tell the story of the development of the discipline, so that the people, expeditions, and publications appear in approximately the correct chronological order, although this means that the fossils themselves will not. However, I will occasionally draw attention to a particular assemblage of animals which coexisted in some ancient environment and thus are now found trapped together in the same cliff or quarry, having achieved fossily fame for their excellent preservation or morphological novelty. Also, I must admit in advance to a bias toward animals rather than plants, and toward backboned animals rather than invertebrates, as well as a more forgivable bias toward life of the last 500 million years because fossils from before that time are immensely rare.

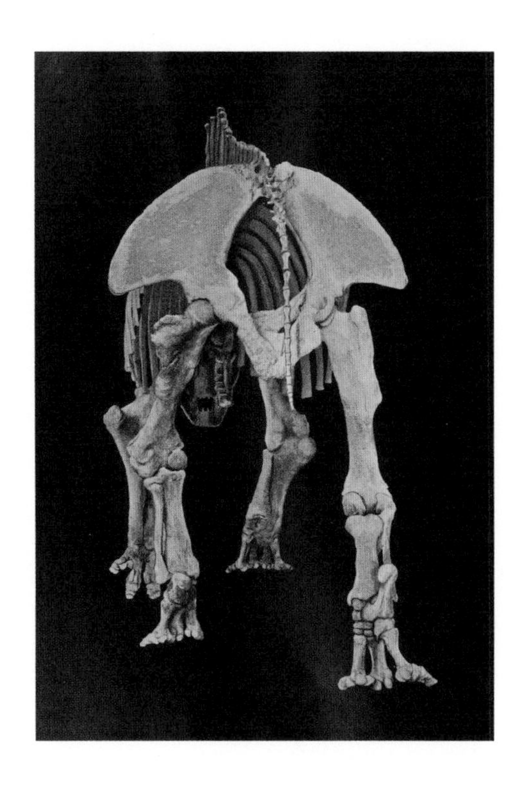

**Henry Fairfield Osborn (1857–1935), *The Titanotheres of Ancient Wyoming, Dakota, and Nebraska*, U.S. Geological Survey, vol. 55, 1929; *Brontotherium*, front view (opposite), rear view (above).**

Indeed, paleontologists are, more than most scientists, limited by what material is available. That one crucial fossil needed to resolve a particular biological conundrum may be out there somewhere in the rocks, but there is no guarantee it will ever be discovered. And increasingly we have realized that fossilization is an irritatingly erratic process. The great majority of animals do not fossilize at all, and those that do may be scattered by flowing water or pulverized by the restless rocks around them. And the story of those few fossils that *do* survive may not be what it seems—early hominin fossils often lack fingers and toes because that is the part big cats liked to chew on, while the ichthyosaurs fossilized apparently giving birth are not captured moments of maternal joy, but records of death in childbirth.

Yet for a story told by the old and the dead, the history of paleontology is a varied and charming one, replete with brilliant, adventurous, and idiosyncratic people. Perhaps there is something about dealing with the old and the dead that keeps one young and alive.

**David Bainbridge, Cambridge, 2021**

# INTRODUCTION: WHEN FOSSILS BECAME LIFE

## ANTIQUITY TO 1800

**"O Time, swift despoiler of created things ... how many changes of states and of circumstances have followed since the wondrous form of this fish died here?"**

Leonardo da Vinci

Many centuries of philosophizing went by before the concept of paleontology as a scientific discipline ever existed. Humans have been stumbling upon the petrified remains of ancient animals since prehistoric times, but the recognition of what those remains actually are took an achingly long time.

The instinctive human reaction to finding vestiges of oversized creatures, or fossilized animals far removed from where those animals' modern relatives now live, has often been to force the fossils to fit human cultures' existing mythologies—or possibly even help form those mythologies. In Japan, fossil shark teeth were seen as evidence of a legendary "heavenly dog"; the bones of giant ancient reptiles probably fed into North American Plains tribes' stories of the thunderbird; and in China fossil bones were often called "dragon bones" (see page 7) and even ground up for use in traditional medicines.

Fossils abounded in the classical world—giant mammals around the basin of the relatively recently inundated Mediterranean, dinosaurs protruding from cliffs along the trade routes into Asia, and mollusc shells almost everywhere, even at the summits of mountains. From as early as the seventh century BCE, we have evidence that the Greeks and their intellectual allies collected and studied fossils

**Below: Robert Plot (1640–1696), *The Natural History of Oxford-Shire*, 1677; Distal femur of a *Megalosaurus*.**

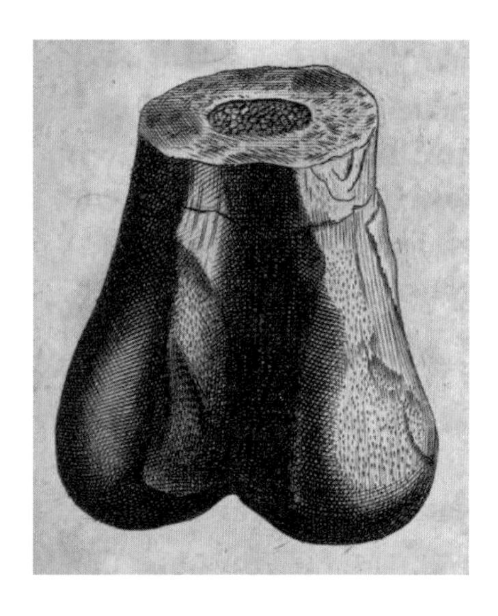

**Opposite: Nicolaus Steno (1638–1686), *Canis Cacharia Dissectum Caput*, 1667; Head of a great white shark.**

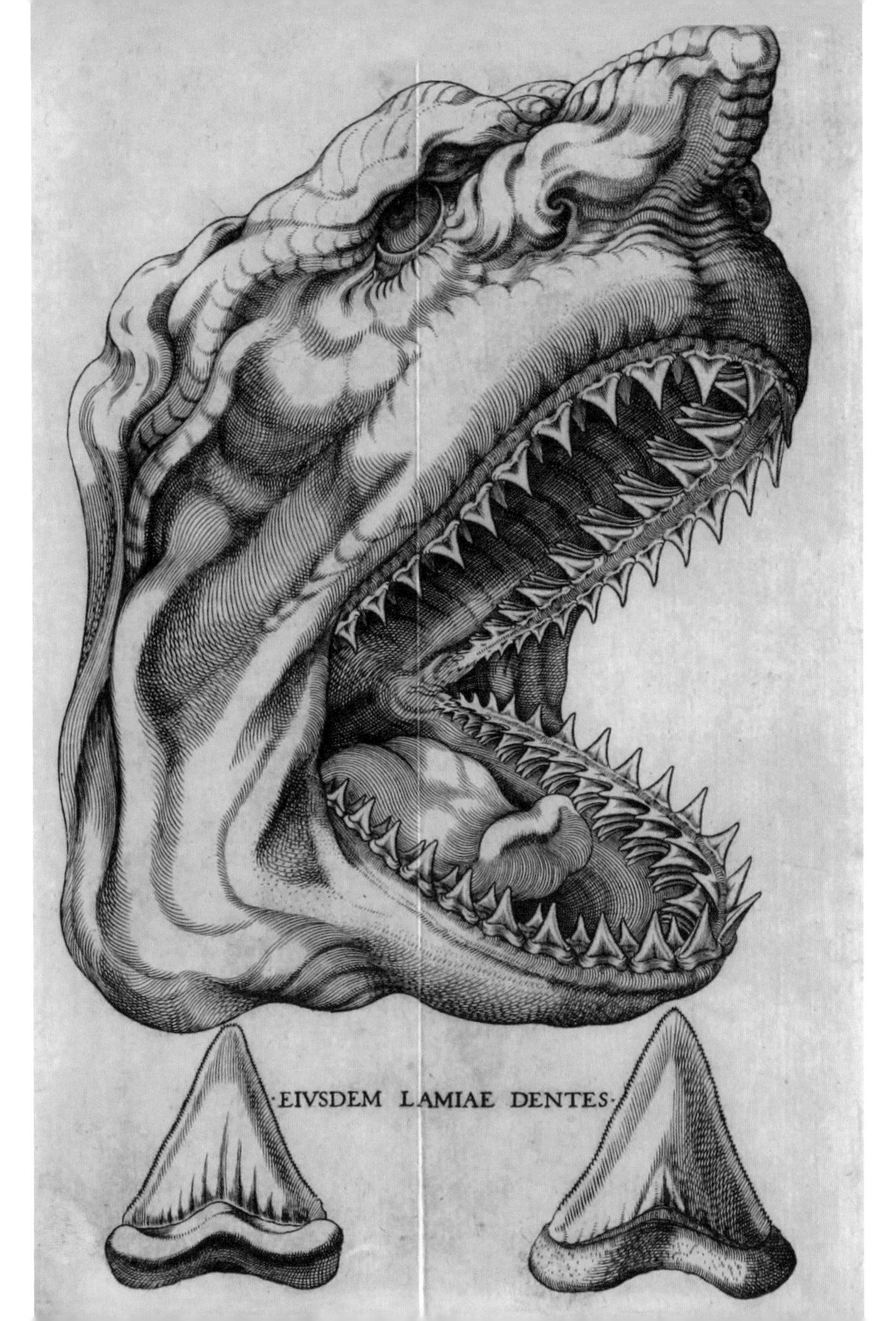
·EIVSDEM LAMIAE DENTES·

and even venerated them in temples, while later one Roman emperor displayed them in his palace's "museum of monsters." Some large fossils, often extinct elephant species, probably contributed to the myths of the Cyclops, the Calydonian Boar, and Geryon the Giant. Possibly the earliest surviving depiction of a fossil is a giant skull painted on a ceramic bowl which recounts the story of the *Ketos Troias*, the Monster of Troy (see page 6). Yet not all classical interpretations of fossils were mythological—as early as 500 BCE, Xenophanes speculated that the strange distribution of fossils might mean the world, or large parts of it which today are dry land, was once covered by an ancient ocean.

This idea that the geography of the Earth can change over time was bolstered when the Persian thinker Abu Rayhan al-Biruni (973–1048) proposed in his treatise on India that fossil evidence suggested that the region had once been under the sea. Indeed, the Islamic Golden Age was crucial in advancing our ideas of ancient life—for example, more than one philosopher proposed theories about the changeability of animal forms similar to the theories of evolution and natural selection formulated in the nineteenth century. And in 1027 the Persian scientist and physician Ibn Sina (980–1037) even sought to explain fossils' apparent stoniness in his *Book of Healing*. Presciently, he stated that fossils are formed when the remains of dead animals and plants are turned to stone by a "petrifying virtue" which seeps from the earth, perhaps during earthquakes or other, slower upheavals.

Five centuries later, the renaissance polymath Leonardo da Vinci (1452–1519) developed remarkable insights into the nature of fossils. Leonardo's love of rock formations is obvious from his art, but he also wrote in his notebooks about discovering ancient footprints, burrows,

**The small feathered theropod dinosaur *Ubirajara jubatus*, discovered in Brazil in 1995. Reconstruction by Luxquine, 2020.**

feces, and bones locked in the rocks of the Tuscan hills. He describes stumbling upon giant bones in a cave, whale bones lodged up in the heights, and one of his diagrams strongly resembles *Paleodictyon*, a 500-million-year-old hexagonal rock pattern, possibly fossilized burrows, which still mystifies paleontologists today. He rejected the prevailing idea that fossils are just *lusi naturae* (games of nature), reflecting the Earth's inherent tendency to create life from stone, and cited evidence of parasitic damage to fossil shells as proving that they were once very much alive. He even proposed that the landscape around us has been formed by immensely slow and prolonged processes, with water as a primary sculpting force.

Soon after, in 1565, among his many key contributions to biology, the Swiss philosopher Conrad Gessner (1516–1565) published his *De Omni Rerum Fossilium* in which he demonstrated the close similarities between fossil and living crabs and sea urchins, and once again questioned how marine species had made their way, in fossil form, so far from the sea.

However, the most important figure in proto-paleontology is the Dane Niels Stensen (1638–1686), more often known as Steno. Steno's cogitations about the Earth and its rocky animal remnants seem to have started early—he worked first at the University of Leiden, in the Netherlands, and having perused nearby fossil collections, he wrote: "Either they have remained there after an ancient flood or because the bed of the seas has slowly changed. On the change of the surface of the earth I plan a book." He soon moved to Florence where he was tasked with creating a "cabinet of curiosities" for the Grand Duke of Tuscany. Famously, in 1666, a great white shark was caught off the coast at Livorno, and it was decided that Steno was the person to whom its severed head should be sent. In his 1667 *Canis Cacharia Dissectum Caput*, Steno stated that modern sharks' teeth are almost identical to the *glossopetrae* or "tongue stones" long claimed to have fallen from the sky. He also made the intellectual leap of suggesting that fossils form when the "corpuscles," which constitute living teeth, are slowly replaced by mineral "corpuscles" to create a solid fossil within solid rock, hence the title of his 1667 *De Solido intra Solidum*.

Remarkably, Steno also developed an entire theory of geology to explain his findings—that the surface of the earth accumulates by the "superposition" of layers of rock laid down when in liquid form, but that these neat, chronologically arranged layers could be subsequently moved or disrupted. In other words, Steno had realized that fossils are trapped

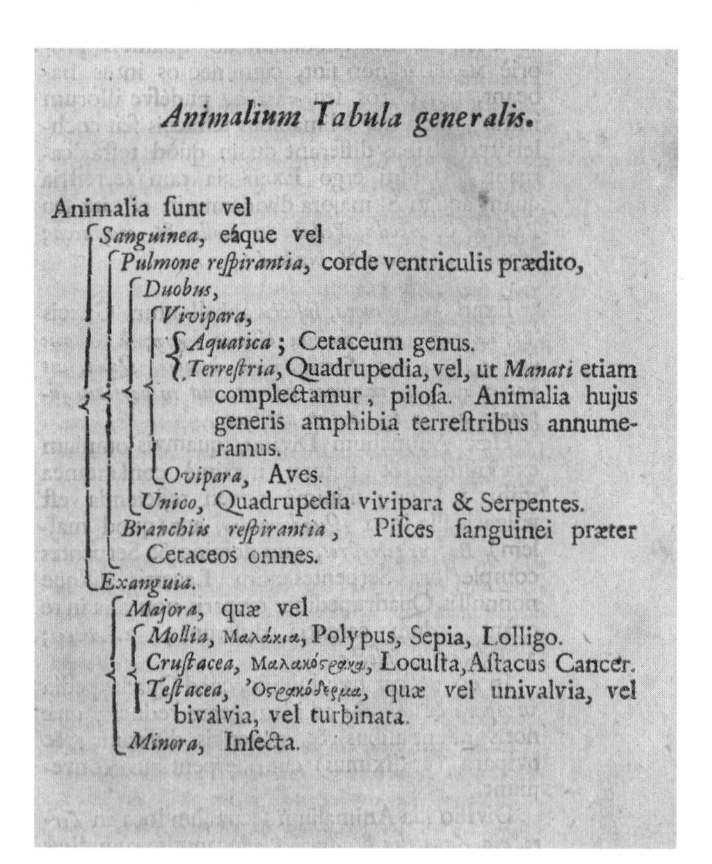

*Animalium Tabula generalis.*

Animalia ſunt vel
- *Sanguinea*, eáque vel
  - *Pulmone reſpirantia*, corde ventriculis prædito,
    - *Duobus*,
      - *Vivipara*,
        - *Aquatica*; Cetaceum genus.
        - *Terreſtria*, Quadrupedia, vel, ut *Manati* etiam complectamur, piloſa. Animalia hujus generis amphibia terreſtribus annumeramus.
      - *Ovipara*, Aves.
    - *Unico*, Quadrupedia vivipara & Serpentes.
  - *Branchiis reſpirantia*, Piſces ſanguinei præter Cetaceos omnes.
- *Exanguia*.
  - *Majora*, quæ vel
    - *Mollia*, Μαλάκια, Polypus, Sepia, Lolligo.
    - *Cruſtacea*, Μαλακόστρακα, Locuſta, Aſtacus Cancer.
    - *Teſtacea*, Ὀστρακόδερμα, quæ vel univalvia, vel bivalvia, vel turbinata.
  - *Minora*, Inſecta.

**John Ray (1627–1705), *Synopsis Methodica Animalium*, 1693; A general table of animals.**

in rocks by what we would now call the processes of sedimentation and eruption, and disturbed by geological folding, faulting, or intrusion. Despite his radical ideas, Steno was careful to fit his theories into an orthodox Biblical chronology, and it is for this reason they were not condemned by the authorities. In fact, Steno soon converted to Catholicism himself, took up orders, and eventually became a bishop.

In the seventeenth century the use of the word "fossil" became more common, a term which rather vaguely means an entity that is "obtained by digging." Also, 1677 saw the first description in the scientific literature—Robert Plot's *The Natural History of Oxford-Shire*—of a dinosaur fossil, although it was not yet recognized as such. The end of a huge thigh bone had been found in that county, which Plot interpreted as belonging to a Roman war elephant or giant of Biblical proportions. A century later, the bone's two bulbous terminal "knuckles" led another savant to name it *Scrotum humanum*, a name which this dinosaur species has only recently been fortunate to have officially changed to *Megalosaurus bucklandi*.

Around this time, the English naturalist John Ray (1627–1705) was taking issue with Biblical chronology, a Christian orthodoxy itself largely fabricated in the previous century. He indicated many fossils are the remnants of animals that lived in primordial seas where now there is land, raised up above the waves by "subterraneous fires and flatuses." He also argued that a Biblical flood would not wash marine fossils *up* mountainsides. In contrast to Ray, his contemporary Robert Hooke (1635–1703), the Curator of Experiments at the newly formed Royal Society, argued that most fossils represented animal types that have now become extinct—"tokens of antiquity," he called them. He studied ammonites, crabs, molluscs, and mammalian teeth, and in his

*Micrographia* of 1665 described petrified wood, which he argued was produced by "petrifying" water percolating minerals through dead tissue.

The last great figure in this brief tour of proto-paleontology is the Scottish farmer, naturalist, and geologist James Hutton (1726–1797), the thinker most responsible for our modern view of the world as an immensely ancient but ever-changing place. Scotland is the ideal place to be a geologist, as there are few places on Earth with such a diverse range of geological forms in such a small area. Hutton was particularly interested by the abrupt boundaries he observed in the Earth's crust where rock layers of different types, arrangements, or orientations meet. It was clear to Hutton that this phenomenon did not suggest that the world was formed in one coherent episode and then remained unchanged forever. Instead, he proposed that the apparently immutable rocks on which we stand are continually being created and destroyed in a "great geological cycle" of eruption, erosion, and sedimentation, followed by crumpling and tilting—slow but vast processes driven by subterranean heat. In his 1788 *Theory of the Earth* he contrasted this ever-changing nature of the Earth with his hypothesis that the *processes* driving this change have remained essentially the same from the origin of the planet to the present day.

**Robert Hooke (1635–1703), *Posthumous Works*, 1705; Snake-stones, or ammonites.**

This "uniformitarianism" has become a central tenet of geology and paleontology, and indeed a subtext for all modern science. The universe changes but its processes do not. It began to dawn on many that the Earth is, in fact, unimaginably old, but has recorded a convenient chronology of ancient life in the layered rocks. And in the first chapter we will see how this realization of immense deep time, including time for animal forms to change, coincided with the first great era of discovery of the petrified remains of those animals.

**Above: Barnum Brown and excavators lifting the encased sacrum of a *Tyrannosaurus rex*, Hell Creek, Montana, 1908.**

**Opposite: Louis Figuier (1819–1894), *La Terre Avant le Déluge*, 1871; Stratigraphic diagram aligned with figures from the Interactive International Chronostratigraphic Chart (2020) of the International Commission of Stratigraphy.**

For reference while reading this book, opposite is an alignment of a late-nineteenth-century stratigraphic diagram with the modern, internationally agreed geological chronological system. By 1871 it was largely accepted that the Earth's crust comprises rock layers laid down over time, with ever more ancient sediments deeper down—although this neat arrangement is often distorted and damaged by erosion, deformation, or eruption. In this diagram all geological "periods" except the most ancient and modern are labeled as they would be today, albeit with different relative durations. The inextricable links between geological science and zoology are evidenced by the fact that the great stratigraphical "eras" are now called the "Paleozoic," "Mesozoic," and "Cenozoic": the "old animals," "middle animals," and "new animals."

Throughout this book, although geological eras will be mentioned, ages will be stated in "millions of years," to allow ease of comparison. The Earth is approximately 3,500 million years old; multicellular fossils date back approximately 600 million years; the non-avian dinosaurs became extinct 66 million years ago; and humans are thought to have diverged from chimpanzees around 6 million years ago.

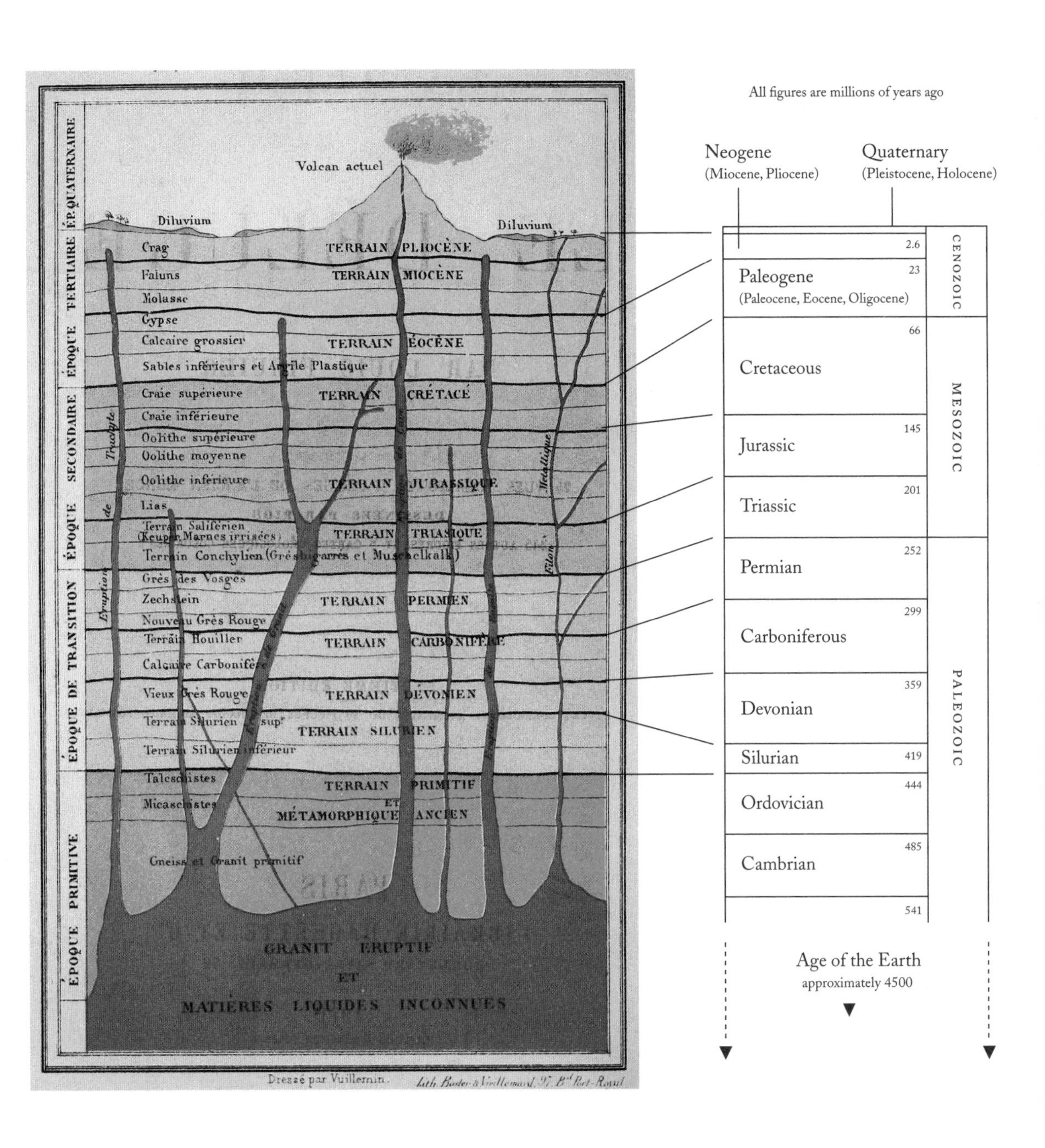
Volcan actuel
Diluvium
Diluvium
ÉP. QUATERNAIRE
ÉPOQUE TERTIAIRE
ÉPOQUE SECONDAIRE
ÉPOQUE DE TRANSITION
ÉPOQUE PRIMITIVE
Crag
TERRAIN PLIOCÈNE
Faluns
TERRAIN MIOCÈNE
Molasse
Gypse
Calcaire grossier
TERRAIN ÉOCÈNE
Sables inférieurs et Argile Plastique
Craie supérieure
TERRAIN CRÉTACÉ
Craie inférieure
Oolithe supérieure
Oolithe moyenne
Oolithe inférieure
TERRAIN JURASSIQUE
Lias
Terrain Salifèrien
(Keuper, Marnes irrisées)
TERRAIN TRIASIQUE
Terrain Conchylien (Grès bigarrés et Muschelkalk)
Grès des Vosges
Zechstein
TERRAIN PERMIEN
Nouveau Grès Rouge
Terrain Houiller
TERRAIN CARBONIFÈRE
Calcaire Carbonifère
Vieux Grès Rouge
TERRAIN DÉVONIEN
Terrain Silurien sup.
TERRAIN SILURIEN
Terrain Silurien inférieur
Talcschistes
TERRAIN PRIMITIF
Micaschistes
ET
MÉTAMORPHIQUE ANCIEN
Gneiss et Granit primitif
GRANIT ÉRUPTIF
ET
MATIÈRES LIQUIDES INCONNUES
Éruption de Trachyte
Filon Métallique
Dressé par Vuillemin.
All figures are millions of years ago
Neogene
(Miocene, Pliocene)
Quaternary
(Pleistocene, Holocene)
2.6
Paleogene
(Paleocene, Eocene, Oligocene)
23
CENOZOIC
66
Cretaceous
MESOZOIC
145
Jurassic
201
Triassic
252
Permian
299
Carboniferous
359
Devonian
Silurian
419
PALEOZOIC
444
Ordovician
485
Cambrian
541
Age of the Earth
approximately 4500

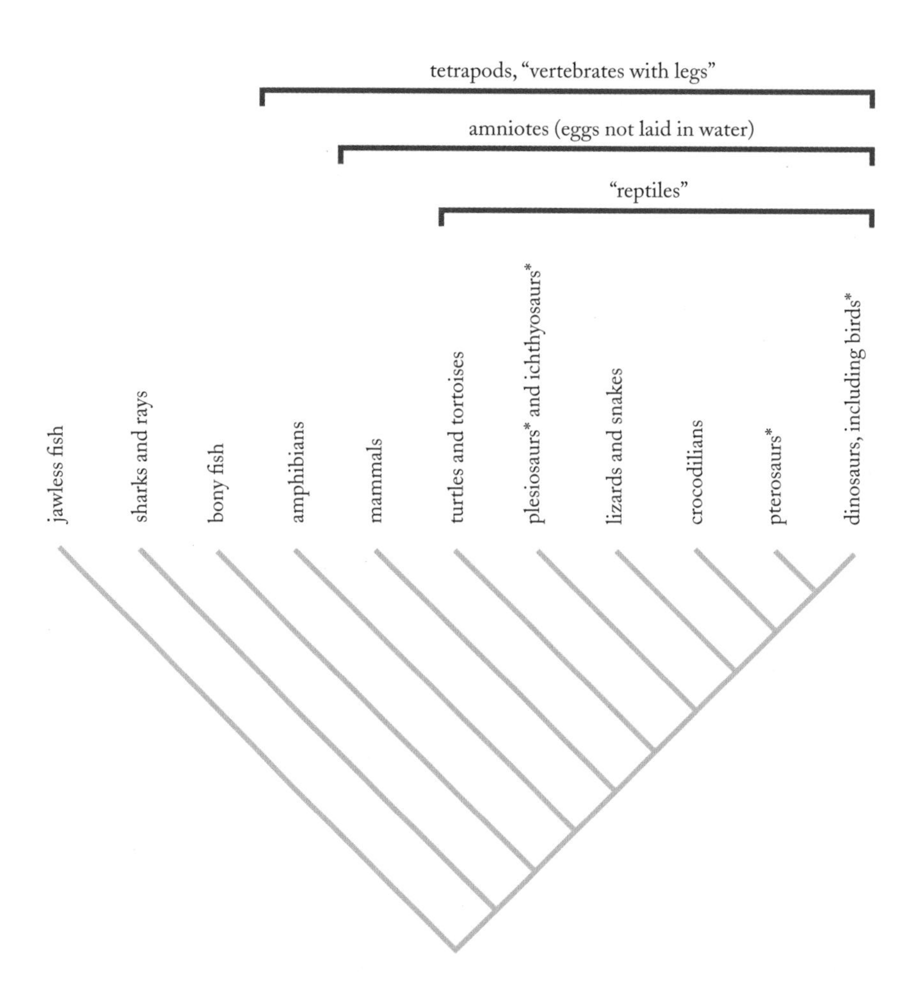


**Above: Illustration by the author; a tree of suggested relationships between some selected vertebrate groups, with asterisks denoting extinct forms.**

Groups connected by shorter branches are more closely related—they diverged more recently—than those connected by longer branches. Any such tree is necessarily tentative, in particular, the relationships of plesiosaurs, ichthyosaurs, and turtles and tortoises remain controversial.

**Opposite: Charles Willson Peale (1741–1827), *The Artist in His Museum*, 1822, Pennsylvania Academy of Fine Arts.**

71
14
15
73
21
33
32

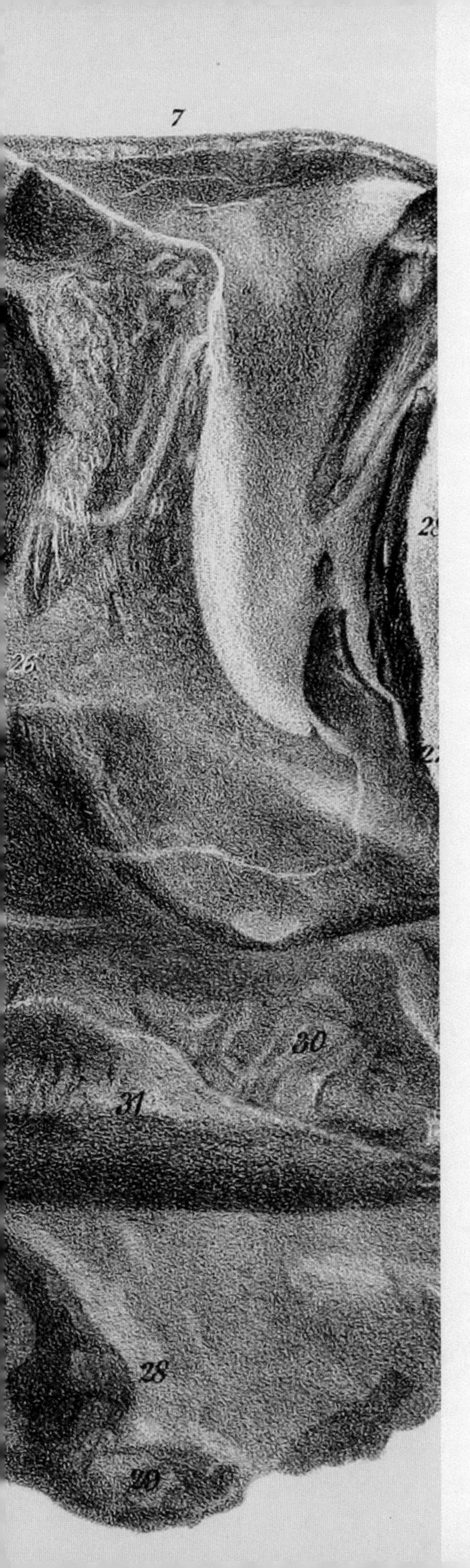


Chapter One

# ORGANIC REMAINS OF A FORMER WORLD

1800–1860

Joseph Dinkel (1806–1891), *Monograph of the Palaeontographical Society, London*, vol. 13, 1861; Holotype skull of *Scelidosaurus harrisonii* from the Lower Lias (early Jurassic), Charmouth, Dorsetshire.

# ORGANIC REMAINS OF A FORMER WORLD

## 1800–1860

**"It is certainly a wonderful instance of divine favour – that this poor, ignorant girl should be so blessed ... she understands more of the science than anyone else in this kingdom."**

Harriet Silvester

It could be argued that modern paleontology started in 1811 when a twelve-year-old girl excavated a fossil on the south coast of England. Mary Anning (1799–1847; see page 52) was born into a poor family who lived in Lyme Regis on what is now called, thanks largely to her, the Jurassic Coast. In fact, the family lived so close to the sea that their first home was destroyed by waves during a storm. Mary was one of just two siblings out of ten who survived childhood—it had been her brother, in fact, who first discovered the fossil of a marine creature that proved to be of such significance in the field of paleontology. Their father had recently died and Mary was trying to continue the family's meager fossil-selling

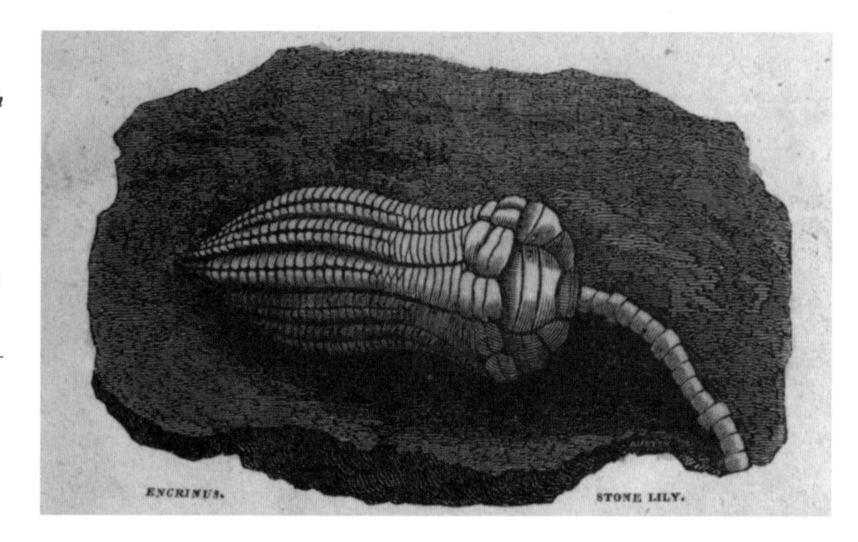


**James Parkinson (1755–1824), *Organic Remains of a Former World: An examination of the mineralized remains of the vegetables and animals of the antediluvian world; generally termed extraneous fossils*, vol. 1, 1811; Fossil of a Crinoid (a type of marine invertebrate related to starfish).**

business, simultaneously teaching herself biology and dissecting animals that had been stranded by the tide. This new fossil turned out to be an ancient, strikingly dolphin-shaped marine reptile, 17ft (5.2m) long, which is now displayed in the Natural History Museum in London. Mary was to make other spectacular discoveries in the future, but it was the otherworldliness of her *Ichthyosaurus* (fish-lizard) which changed science. It was a striking creature, obviously an extinct one, and it was discovered as attitudes to the history of life were changing.

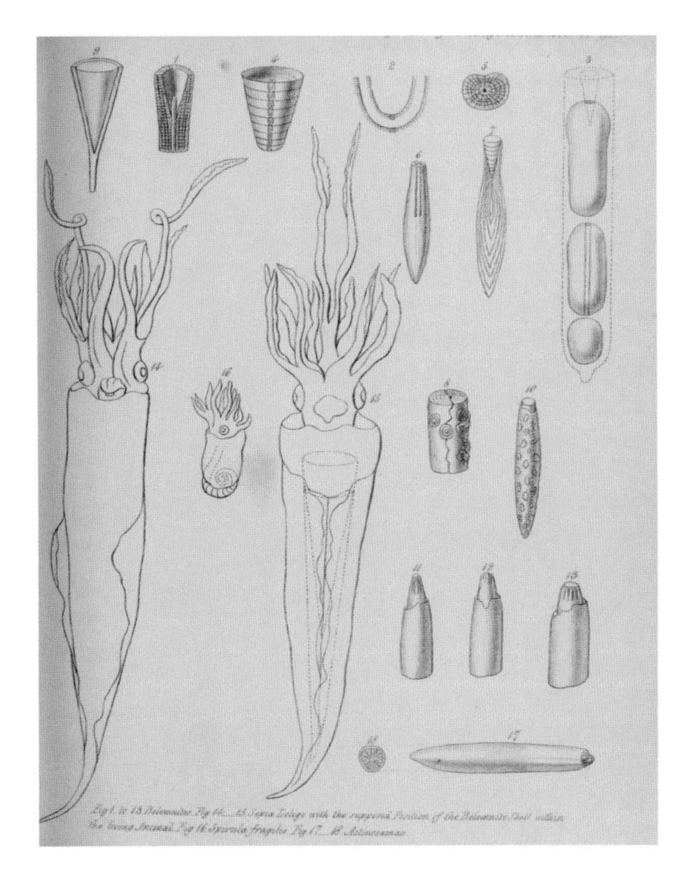

**Belemnites, from J. S. Miller, "Observations on Belemnites," *Transactions of the Geological Society, New Series*, series 2, vol. 2, plate IX, 1826.**

The idea that animal types might sometimes be extinguished was not new. Indeed, in the previous decade scientists had started to classify fossils into those which resembled existing species living nearby, those which resembled existing species now living far distant, and those which seemed like nothing now alive. Also, Jean Baptiste Lamarck (1744–1829) of the Muséum national d'Histoire naturelle in Paris had become an influential proponent of the idea that animal species change and split over time in response to their environment. Thanks to his 1809 *Philosophie Zoologique*, the concepts of extinction and evolution were very much in the contemporary scientific mind.

Yet paradoxically George Cuvier (1769–1832), one of the driving forces behind early-nineteenth-century paleontology, was very much against the nascent theory of evolution. Also working in Paris, he agreed that periodic cataclysmic extinction events were important in the history of animal life, but not gradual change in individual species. Most importantly, though, Cuvier began to apply the science of comparative anatomy to fossils—as early as 1800 he had argued that fossils are extinct species of animals and began to characterize them by comparison with living forms. He studied a giant "beast from Paraguay" (actually Argentina) which he realized was an enormous sloth, compared mammoths and mastodons from the USA and Russia to modern

**Thomas Landseer (1795–1880), *The Voyage of the Beagle*, 1839; H.M.S. *Beagle* laid ashore, Santa Cruz River, April 16, 1834.**

elephants, and also characterized a variety of mammals excavated in nearby Montmartre. In addition, he described and named a "ptérodactyle," which he characterized as "un reptile volant." This, and the "Beast of Maastricht" (see page 32) which he identified as a giant marine reptile, led him, in his 1812 *Recherches Sur Les Ossemens Fossiles de Quadrupèdes*, to speculate that the Earth was once dominated by reptiles. And Cuvier's name appears again and again in the early story of paleontology, touring the fossil collections of Europe to help identify specimens: certainly he was excited by Mary Anning's discovery.

Cuvier also examined fossils unearthed by William Buckland (1754–1856). Buckland had discovered giant bones in Stonesfield quarry, in Oxfordshire, in 1815, which Cuvier realized were those of a giant lizard, and indeed they represent the first scientifically described dinosaur, *Megalosaurus*. In fact, it is somewhat surprising that so much of the early history of dinosaur hunting occurred in Britain, as it actually possesses a rather meager selection of specimens. Buckland later discovered the "Kirkland Hyaena Den" in a quarry in Yorkshire (see page 40), littered with the bones of animals related to modern elephants, tigers, hippos, and, of course, hyenas, along with fossilized hyena feces. He realized the entrance to the cave was too narrow for pachyderms to be washed in by a Biblical deluge, and argued that they were instead carried there, in pieces, by its resident hyenas in some far more ancient epoch (probably 120 thousand years ago, in fact). The "den" soon became famous, and its discovery was instrumental in changing public perceptions of the age of the Earth.

The twin rising stars of geology and paleontology were very much aligned in those early years. In 1816–19, the English surveyor William Smith (1769–1839; see page 34) published *Strata Identified by Organized Fossils*, in which he carefully recorded the fossils unearthed from sediments across Britain and argued that each "stratum" or rock layer contains entombed remains of its own characteristic, distinctive fauna. He also discussed whether gaps in the fossil record simply reflect its incompleteness or are instead signs of ancient extinction events. Whatever the case, fossils could now be used to assign a "date" to strata, and conversely strata could be used to "date" fossils. Indeed, this process was later developed to an astounding level of accuracy in the 1850s by Friedrich Quenstedt and Albert Oppel working in the Alps of southern Germany—they discovered that ammonites (see page 62) make perfect fossil timepieces as they are abundant, widespread, and evolve rapidly into ever-changing forms, each restricted to precise time windows that are shorter than one million years.

**Thomas Sopwith (1803–1879), *William Buckland fossiling at Cwm Idwal, Snowdonia*. Pen and ink drawing, October 1841.**

In 1822, inspired by Mary Anning, the Sussex doctor Gideon Mantell (1790–1852) discovered remains of the iconic dinosaur *Iguanodon*. Mantell had been studying giant bones from nearby quarries for some time but it was probably his wife Mary Ann who, while out riding, spotted in wayside gravel the fossil teeth and "horns" that were soon attributed to a great 60-ft (18-m) beast. There was much debate about the nature of the creature—Buckland thought the remains were those of a fish—and controversy too, because the teeth suggested a herbivorous habit unlike *Megalosaurus* or any large modern-day reptile. Famously, one conical fragment was initially reconstructed as a horn atop the animal's nose, but is now thought to have been a sharp, defensive thumb-spike. Later, Mantell studied a more complete skeleton in a large slab blasted from a Kent quarry and realized that *Iguanodon*'s proportions were those of a bipedal animal—something then unknown among large reptiles of any era. He also discovered a second dinosaur in 1833, the spiky, armored *Hylaeosaurus*, although this rare and unusual species remains something of a conundrum for paleontologists, unlike *Iguanodon* which we now know to be just one of a diverse, abundant family of dinosaurs.

Evidence of enormous extinct beasts was also being unearthed on the other side of the Atlantic. In the early 1830s Richard Harlan (1796–1843), professor of comparative anatomy at the Philadelphia Museum, was sent some gigantic bones exhumed in Louisiana which he thought represented a huge, ancient sea reptile, and he gave it the name *Basilosaurus*. However, later examination showed that its teeth were more like those of modern whales—although the *saurus* (lizard) epithet has survived to the present day. Another scientist working in the United States was Swiss-born Louis Agassiz (1807–1873), whose studies of fossil fishes extended the clock of life even further back in time, and who was also a great proponent of the idea that the Earth underwent dramatic ice ages in its past.

An overarching, and some might say overbearing, figure in nineteenth-century paleontology was Richard Owen (1804–1892), the thoroughly unlikeable Conservator of the Hunterian Museum of the Royal College of Surgeons and eventual instigator of the founding of London's Natural History Museum. Owen was undoubtedly the premier comparative anatomist of his day, and was responsible for the correct identification of many fossils, although his career was beset by allegations that he often claimed credit which was due to others. In 1842, he conferred the evocative collective term *dinosaur* (terrible lizard) on *Megalosaurus*, *Iguanodon*, and *Hylaeosaurus*—Owen, for all his faults, was the first to realize that these disparate species share defining characteristics not found in ancient flying or marine reptiles. It also does not help Owen's cause that he so aggressively disputed Mantell's correct theory that *Iguanodon* was bipedal.

A permanent monument to all the conflict that plagued Victorian paleontological circles survives in the Crystal Palace Dinosaurs (see page 65). They are giant concrete models of the ancient reptiles then known, unveiled in 1854 for the relocation of the Great Exhibition's giant glasshouse to south London. Mantell was passed over as designer in favor of the natural history artist Benjamin Waterhouse Hawkins (1807–1894), and under Owen's guidance the great beasts became bulky, sprawling behemoths, all very much quadrupedal. One thing Hawkins did do, however, was give the beasts hides with fine scales, a feature confirmed by subsequent rare discoveries of fossil dinosaur skin.

This chapter on the first era of scientific paleontology is bookended by the two great biological theories which influenced all future thinking on the subject—evolution in 1809 and natural selection in 1858. The theory of natural selection was proposed simultaneously in a single paper by Charles Darwin (1809–1882) and Alfred Russel Wallace (1823–1913): *On the Tendency of Species to form Varieties; and on the Perpetuation of Varieties*

**Alfred Russel Wallace (1823–1913), *The Geographical Distribution of Animals*, 1876; A Brazilian forest, with characteristic mammalia.**

**Opposite: Richard Owen (1804–1892), *Memoirs on the Extinct Wingless Birds of New Zealand*, vol. 2, 1879; Richard Owen and a moa skeleton.**

*and Species by Natural Means of Selection*. Darwin had been naturalist on the famous voyage of H.M.S. *Beagle* between 1831 and 1836 and spent much of his time fossil hunting. He sent many large fossil bones back to London which Owen identified as huge armored armadillos, giant sloths, and *Toxodon platensis*, a rodent-toothed, hippo-like creature unique to South America. At the time Darwin wrote in his notebooks of his confusion about how the world might be repopulated by animals in the aftermath of extinction events. And, although he discussed fossils little in his 1859 *On the Origin of Species*, we know his emerging theory of natural selection was partly driven by the evidence of his own eyes that extinct South American species had been succeeded by new animals with "a close relationship" to them.

Wallace's early interests had been more focused on existing species and their geographical distribution, but in the years after the *Tendency of Species* paper, he was sometimes more willing than Darwin to attack thorny problems created by the upheaval in science caused by their theory. Certainly his unrivaled expertise in ornithology led him to become an early proponent of the idea that birds are descended from dinosaurs—birds are their only living descendants, in fact. Thirty years on, he studied the ultimate "missing link" fossil *Archaeopteryx* (see page 86) and persuasively argued that dinosaurs were "reptiles which in some respects approach birds"—then a controversial theory, but now generally accepted.

**Below: Johann Jakob Scheuchzer (1672–1733), *Homo Diluvii Testis* (Man, Witness of the Deluge), as described in *Lithographia Helvetica*, 1726; A giant salamander *Andrias scheuchzeri* found in Öhningen, Germany.**

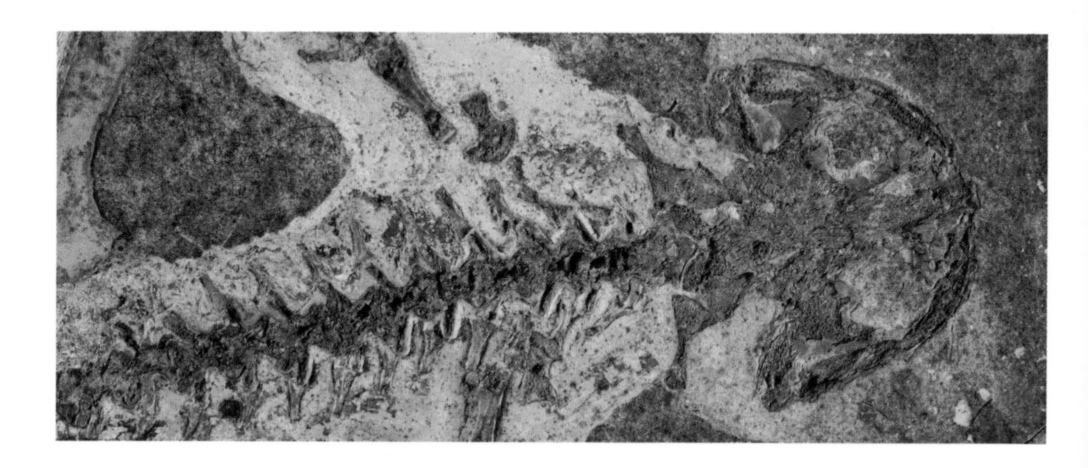

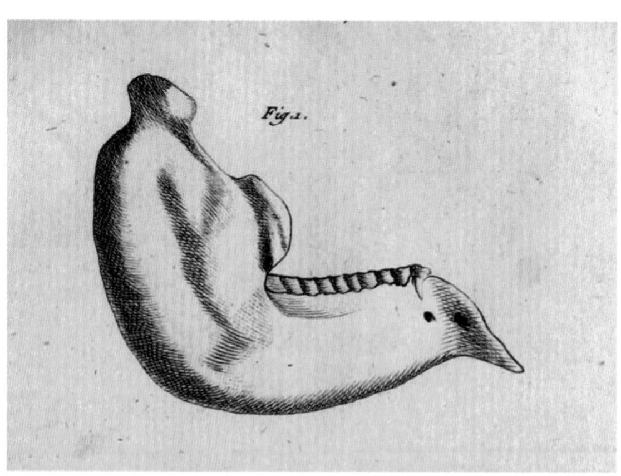


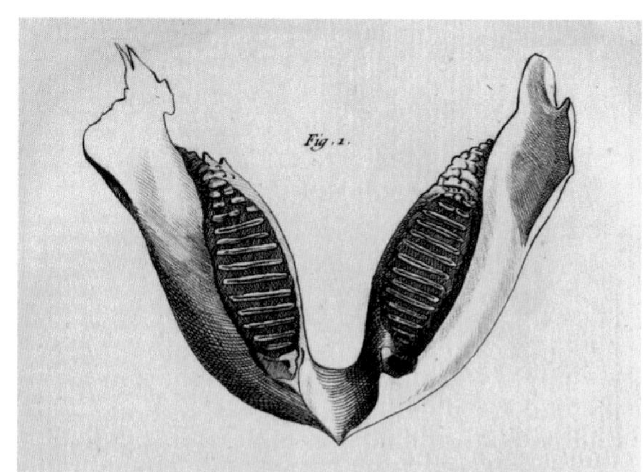


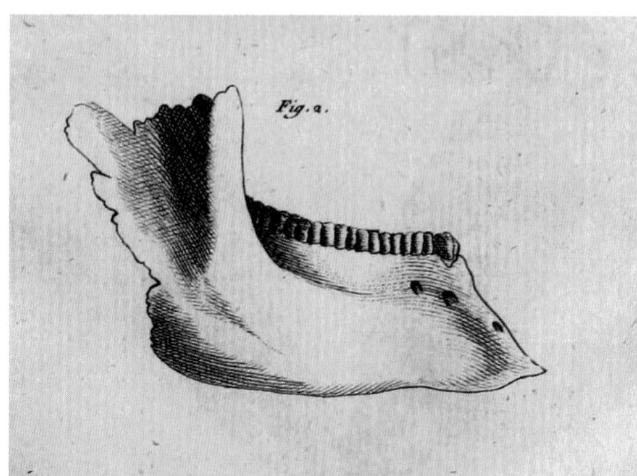


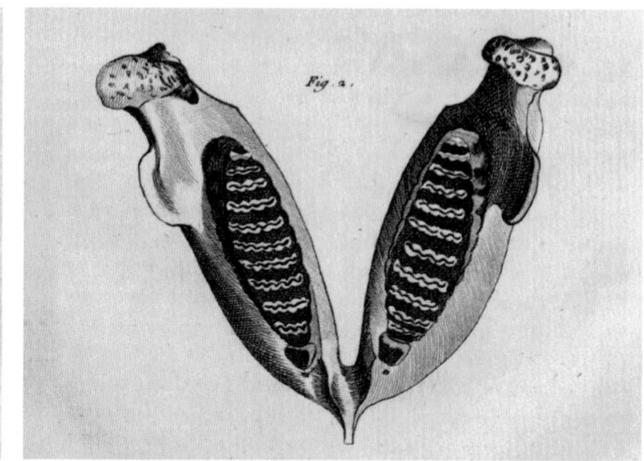


**George Cuvier (1769–1832), *Mémoires de l'Institut des Sciences et Arts*, vol. II, 1798; The jaws of an Indian elephant (top left, bottom right) and a mammoth (bottom left, top right).**

Cuvier argued strongly that fossils are the remains of ancient animals, and was also among the first scientists to apply the new science of comparative anatomy to fossilized specimens. By drawing comparisons between living species of elephants and fossils from the United States and Russia, he demonstrated that the fossil record contains animals similar enough to modern forms to be recognizable, but different enough to be classified as distinct species—and extinct species at that.

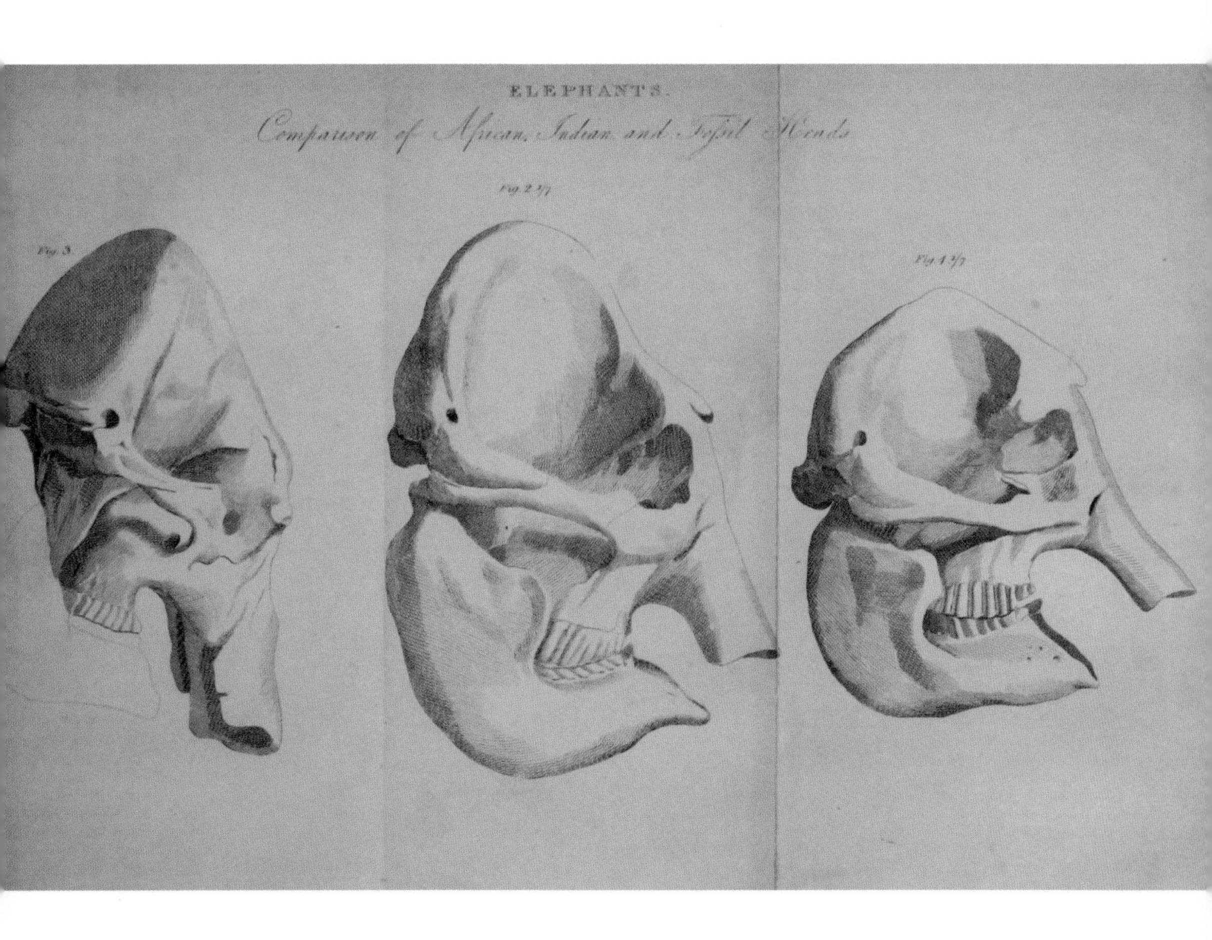


George Cuvier (1769–1832), *Le Règne Animal: Distribué D'Après Son Organisation,* 1816; Elephants: comparison of African, Indian, and fossil heads.

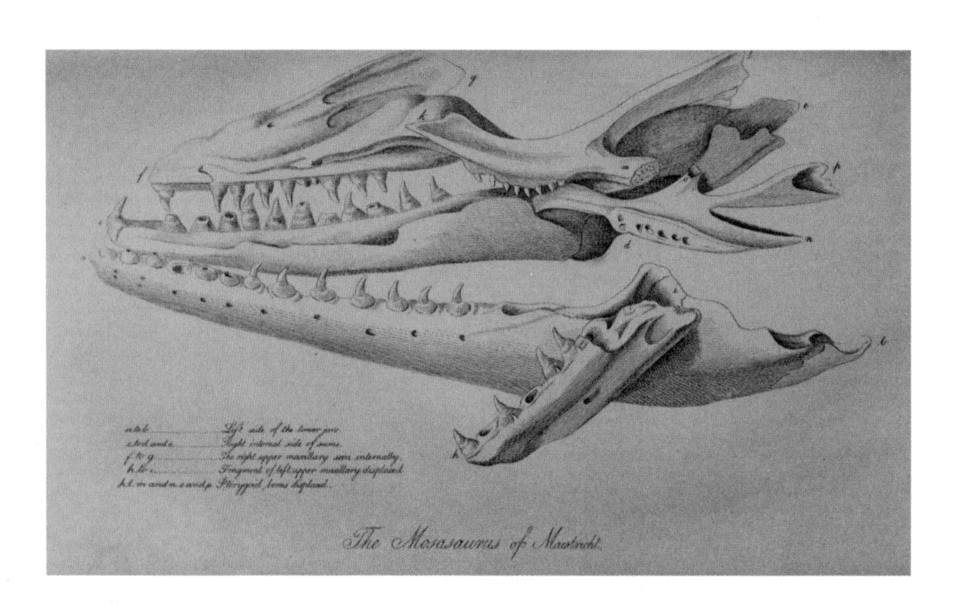


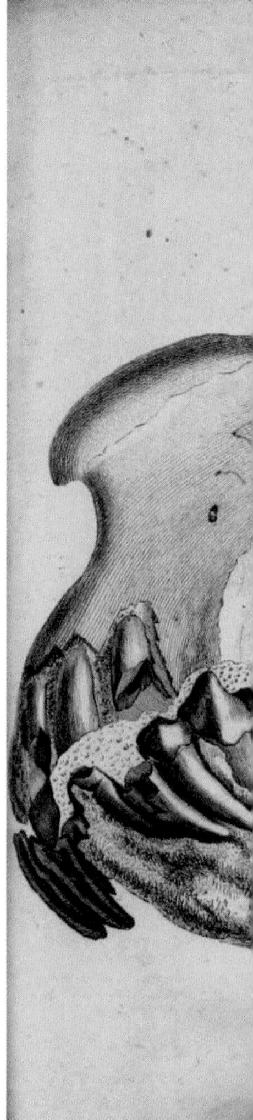

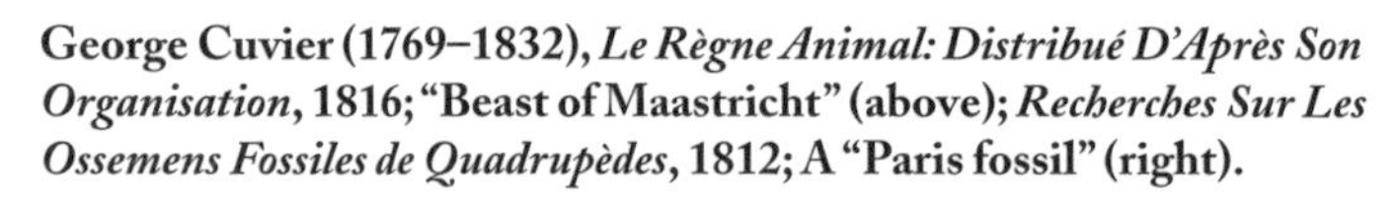

**George Cuvier (1769–1832), *Le Règne Animal: Distribué D'Après Son Organisation*, 1816; "Beast of Maastricht" (above); *Recherches Sur Les Ossemens Fossiles de Quadrupèdes*, 1812; A "Paris fossil" (right).**

Discovered in 1764 in the hill of St Pietersberg, near Maastricht in the Netherlands, the "Beast" is actually a Mosasaur, one of a group of ancient aquatic reptiles probably more closely related to lizards and snakes than dinosaurs, despite their large size—up to 60ft (18m). Initially assumed to be a whale, it was Cuvier who first realized the animal's similarities to modern lizards, thus establishing that it is entirely unlike any known living species. We now know that mosasaurs lived from 80 million years ago to the late-Cretaceous extinction event 66 million years ago. (In contrast, the "Paris fossil" is clearly identifiable as a mammal, due to its multi-rooted teeth.)

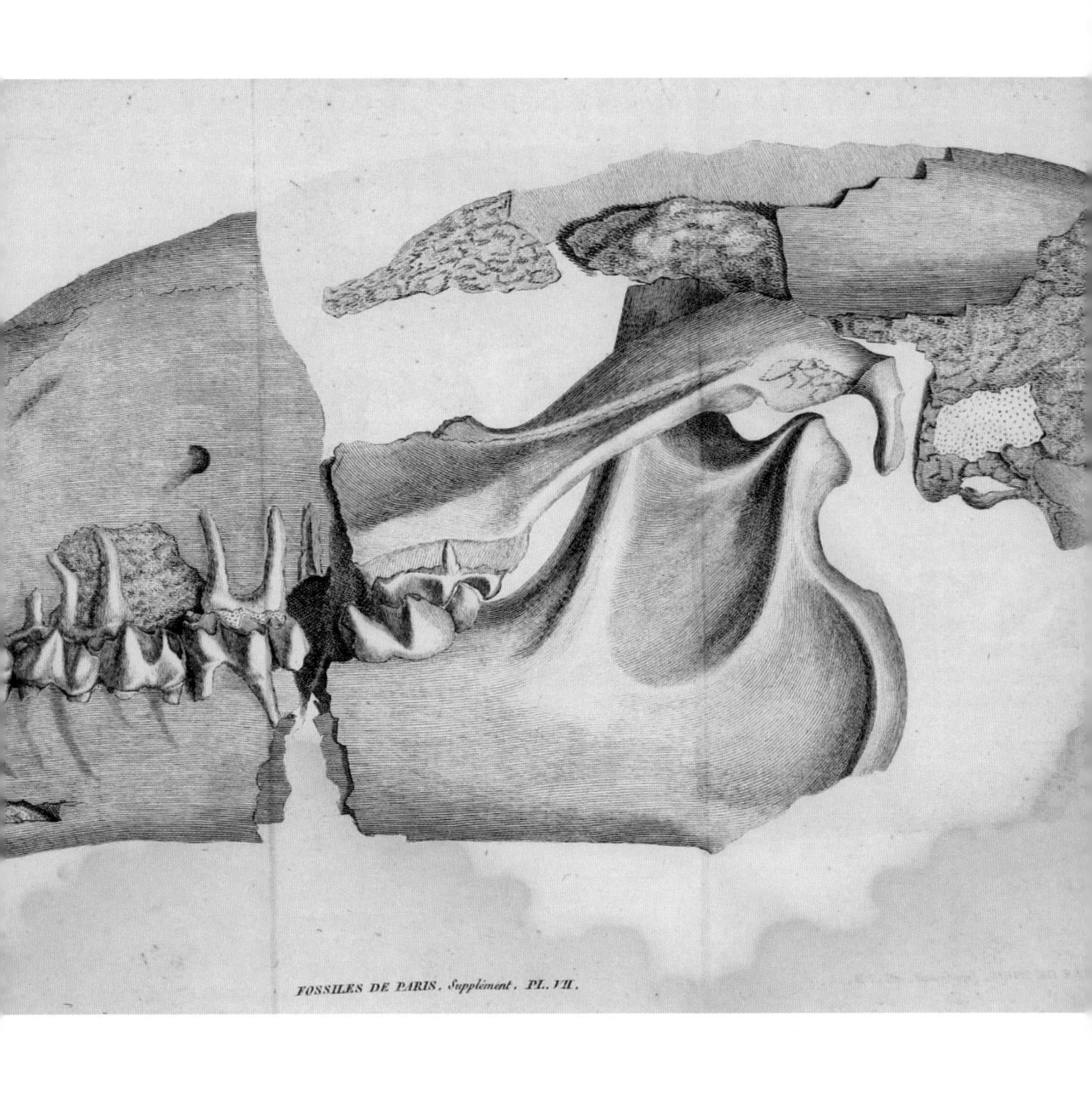
FOSSILES DE PARIS. Supplément. PL. VII.

# WILLIAM SMITH (1769–1839)

## A Layered Chronology of Fossils

As his nickname suggests, the English surveyor William "Strata" Smith became synonymous with our modern scientific realization that many rocks are produced by the gradual deposition of sedimentary rocks in layers, or "strata." In particular, he realized that each stratum contains its own distinctive fossil species.

Born in Oxfordshire to a poor family, Smith spent much of his life in Yorkshire, but wherever he traveled in Great Britain he was struck by the geological diversity conveniently crammed into that small island. There was a real practical and financial imperative for geological surveying in nineteenth-century Britain, as this first industrialized nation became internally connected by new canals and railways. Indeed, Smith's decision to establish his own surveying company was to change his life, and the future of science.

In one of the most productive flurries of scientific publication of all time, Smith published the first detailed, small-scale, modern-looking geological map as his 1815 *A Delineation of the Strata of England and Wales, with Part of Scotland*, and the first systematic matching of sedimentary

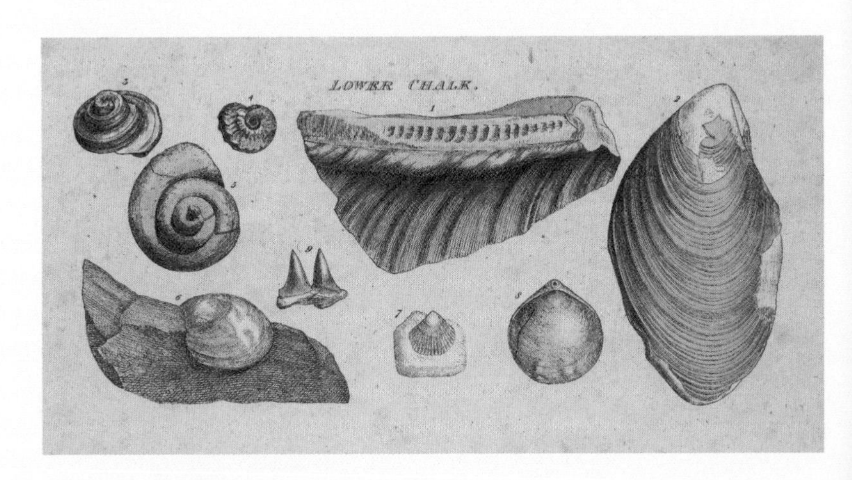


**William Smith (1769–1839), *Strata Identified by Organized Fossils*, 1816–19; Fossils typically found in the lower chalk outcrops on the hills above Bath (right) and in clay over the Upper Oolite, just south of Bath (see overleaf).**

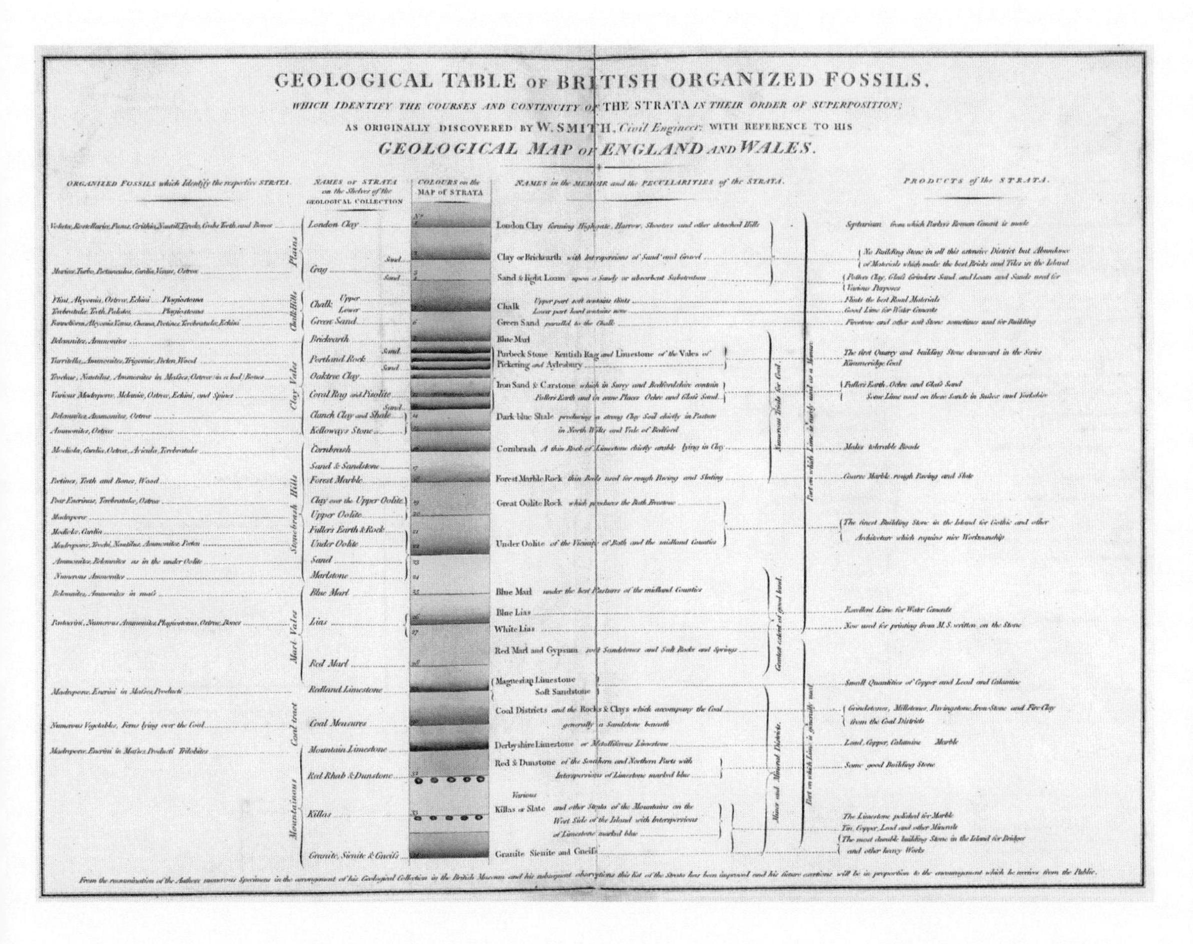


rocks with their entombed fossils in the 1816–19 *Strata Identified by Organized Fossils* and 1817 *Geological Table of British Organized Fossils*.

Smith was most interested in using the continually changing array of fossils in sediments to provide a geological chronology—his "Principle of Faunal Succession"—but from a paleontologist's viewpoint the converse ability to arrange extinct animal forms in an accurate vertical narrative was just as important.

The publication of Smith's beautiful, colorful diagrams placed a tremendous strain on his meager financial means, and he was to die penniless, having sold his own geological and fossil collections to fund his work. Many times, credit for his work was stolen by others and only relatively recently has it become clear that by discovering the layered rocky record of the Earth's past, William "Strata" Smith really did live up to the cliché—one person who changed the way we view the world, and all that lives and has lived in it.

**William Smith (1769–1839), *Geological Table of British Organized Fossils*, 1817.**

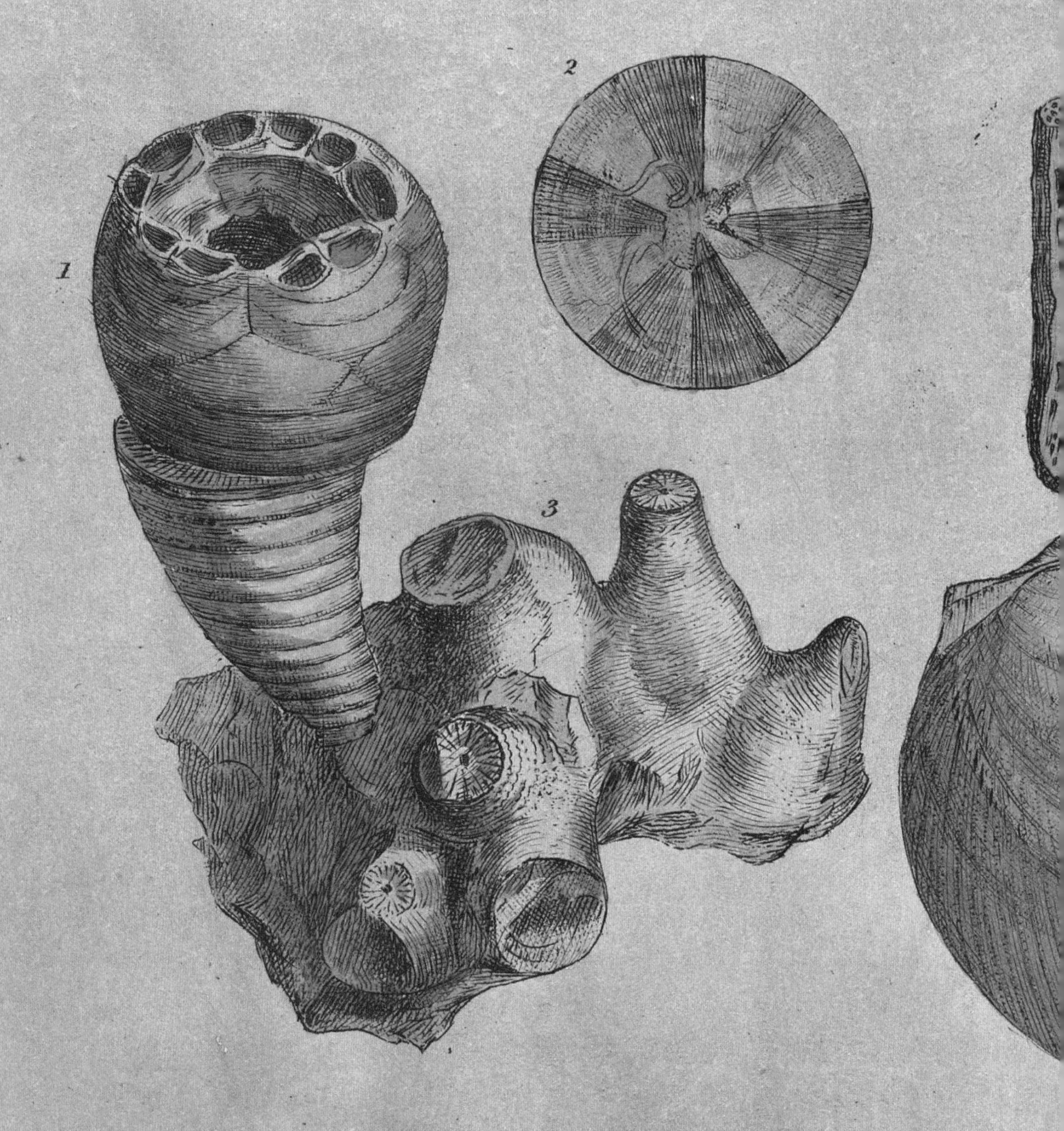


*1.2.3. Pear Encrinus, 2. The Clavicle,*

*3. The Root and Stems,*

*4. Tubipora,*

*5. Millepora,*

*6. Chama cr*

*7. Plagiostor*

R OOLITE.

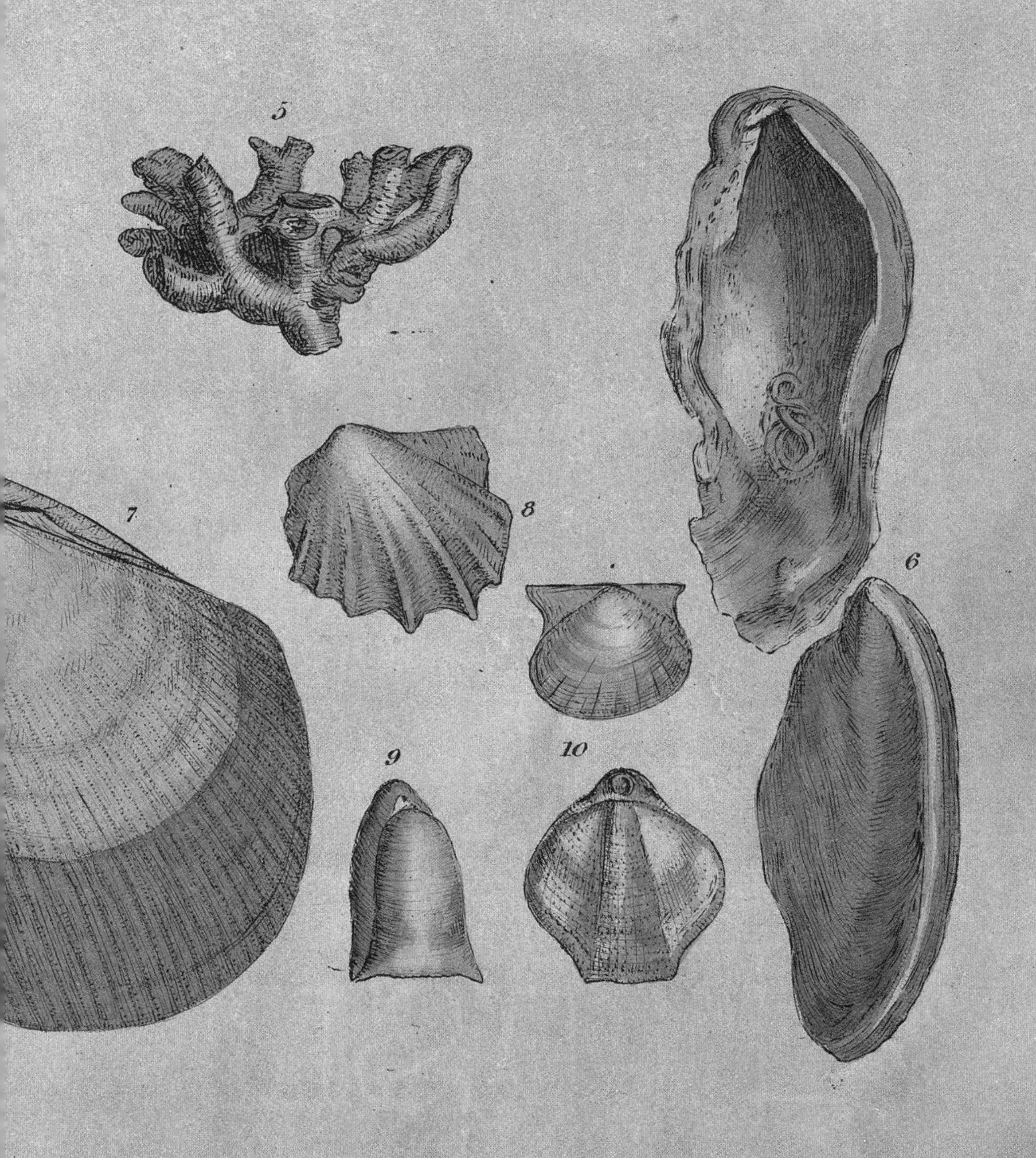


trat. Syst. P. 80.

8. *Avicula costata*, Strat. Syst. P. 81.

9. *Terebratula digona*, M.C. 96.

10. *Terebratula reticulata*, Strat. Syst. P. 83.

**James Parkinson (1755–1824),** ***Organic Remains of a Former World: An examination of the mineralized remains of the vegetables and animals of the antediluvian world; generally termed extraneous fossils*****, vol. 1, 1808; Landscape with ammonite.**

James Parkinson was a surgeon and apothecary who indulged an interest in geology in his spare time, leading eventually to the publication of a three-volume work on life on Earth, past and present, including a discussion of the nature of fossils and fossilization. Few images in paleontology are as evocative as this tiny ammonite shell, cast up on a beach in some unknown epoch.

**Hugh O'Neill (1784–1824), Lithograph of an *Ichthyosaurus chiroligostinus* (now *Temnodontosaurus platyodon*), circa 1820.**

Much of the first ichthyosaur skeleton excavated by Mary Anning (see page 52) has sadly been lost, but more were to follow from the Dorset coast near Lyme Regis, including this spectacular specimen. The animal's modern name contains two instances of the syllable *-don* which means "tooth"—a common element of many fossils' scientific names, since teeth are usually the most durable and well-preserved structures in the body.

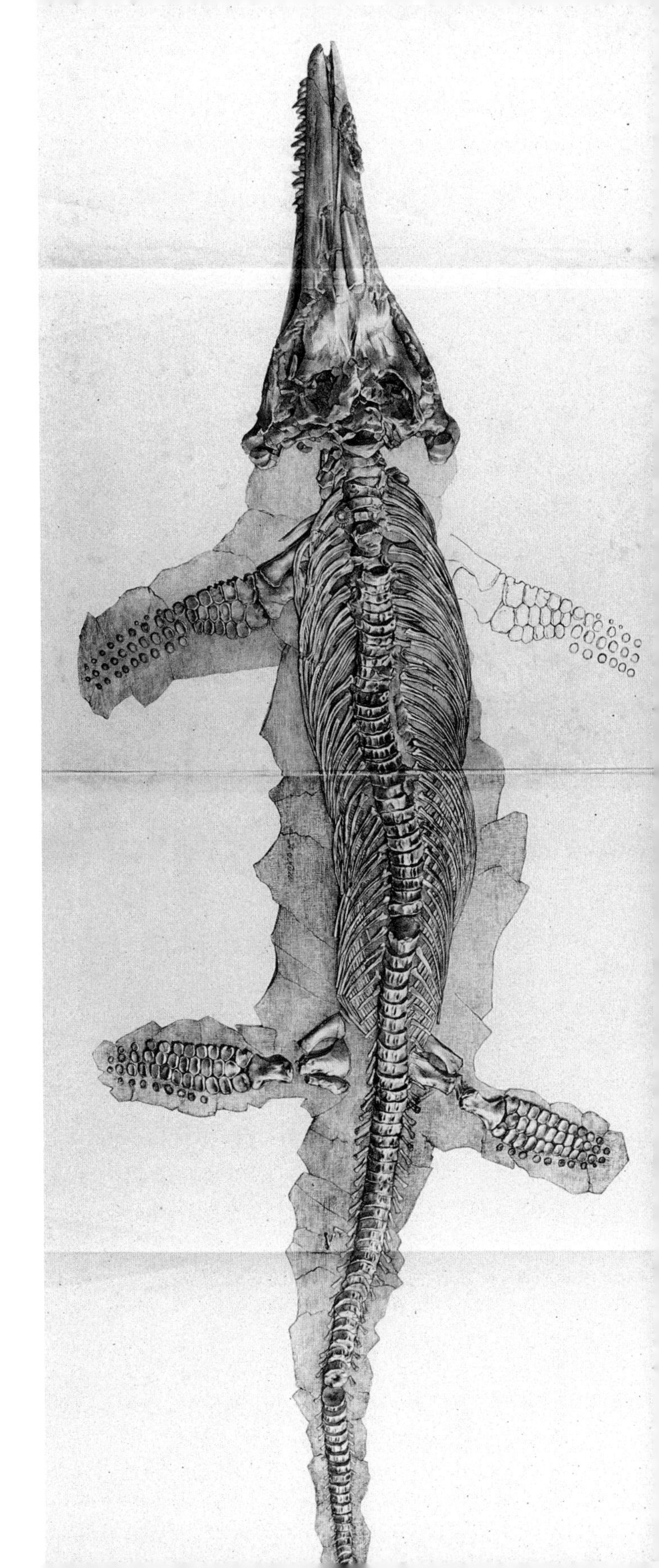

**William Daniel Conybeare (1787–1857),** ***Verses On the Hyaena's Den at Kirkdale near Kirby Moorside in Yorkshire*****, 1822; William Buckland in the Kirkdale Hyaena Den.**

The English geologist and Anglican minister William Buckland wrote and spoke extensively and flamboyantly about a newly discovered Yorkshire cave he was called to investigate. The cave contained the bones of many animals, large and small, familiar and exotic—including the hyenas which were assumed to have dragged fragments of its larger inmates piecemeal through the narrow entrance. Unfortunately, the hyenas were long dead by Buckland's appearance on the scene, contrary to what this fanciful image might have the viewer believe. Buckland soon realized that a Biblical deluge could not account for the presence of remains of such large creatures in a cave with such a small opening, and his famous discovery was to change how the public thought about the Earth's past and the historicity of the Bible. Always the eccentric, Buckland was particularly interested in fossil feces, or "coprolites," and even had some inlaid into a coffee table at his home. In the nineteenth century it was realized that coprolites could be processed into fertilizer, and one of the factories where this processing was done has given its name to the only "Coprolite Street" in the world, in Ipswich in Suffolk.

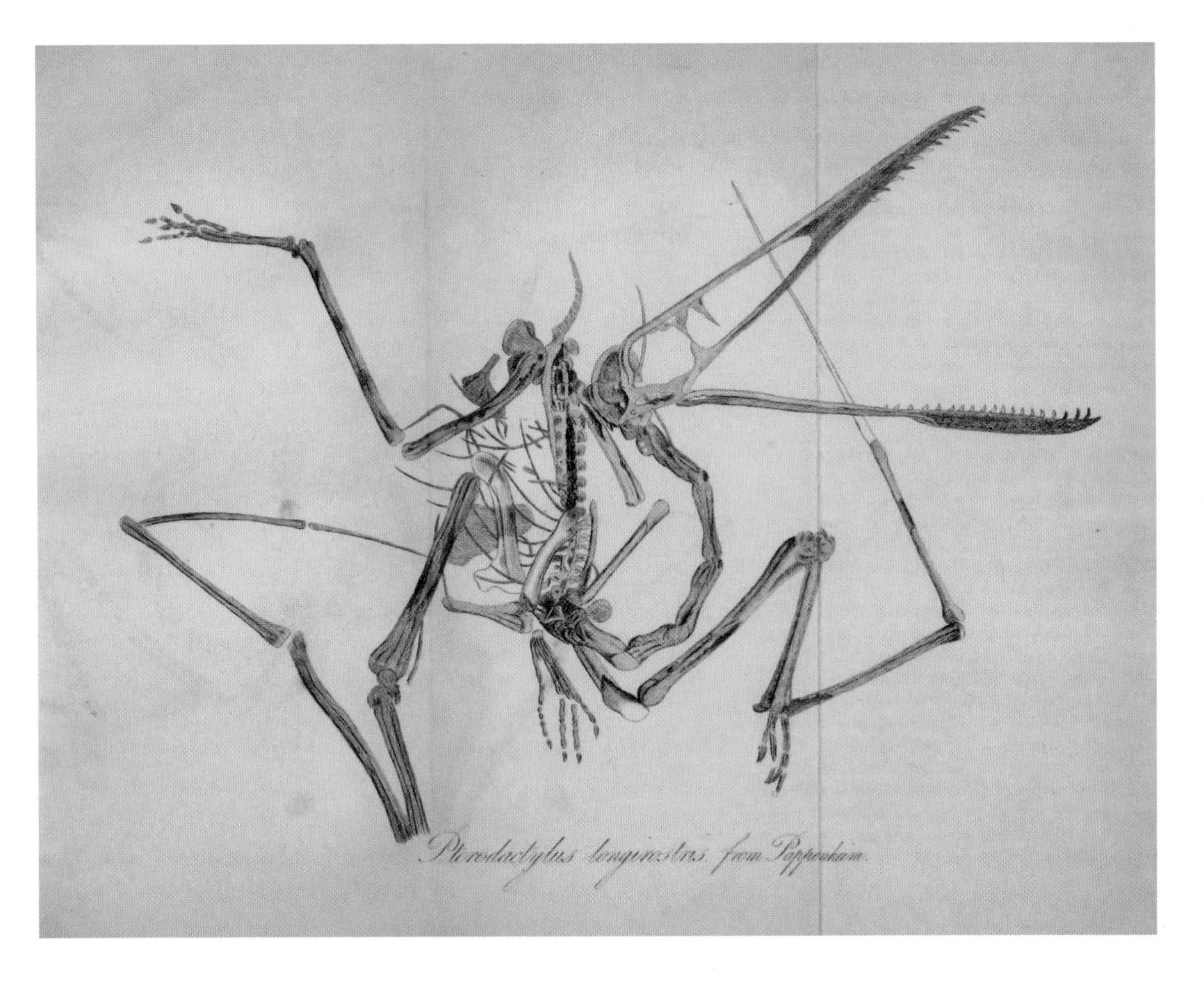


**George Cuvier (1769–1832), *Le Règne Animal: Distribué D'Après Son Organisation*, 1816; Pterodactyl skeleton.**

Within a surprisingly short time in the early nineteenth century, a surge of paleontological research uncovered representatives of many of the major groups of Mesozoic Era (252–66 million years ago) creatures that we know today. These include the marine ichthyosaurs and plesiosaurs, the marine mosasaurs related to lizards and snakes, dinosaurs themselves, and their relatives the pterosaurs. *Pterosaurs* (wing-lizards) appear abruptly in the fossil record, with few intermediate forms linking them to related groups, so their origins remain something of a mystery. However, they clearly represent the first time powered flight evolved in vertebrates—the only other times being birds and bats. Before Cuvier, however, their fragile forms were thought to reflect an aquatic way of life.

**Henry de la Bèche (1796–1855), *Duria Antiquior, or a More Ancient Dorset*, 1830.**

De la Bèche, himself an esteemed geologist and paleontologist, painted this watercolor in large part as a homage to Mary Anning's discoveries. Like many of his depictions, it is a somewhat humorous presentation of an ancient English coastline positively brimming with endearingly strange and savage life.

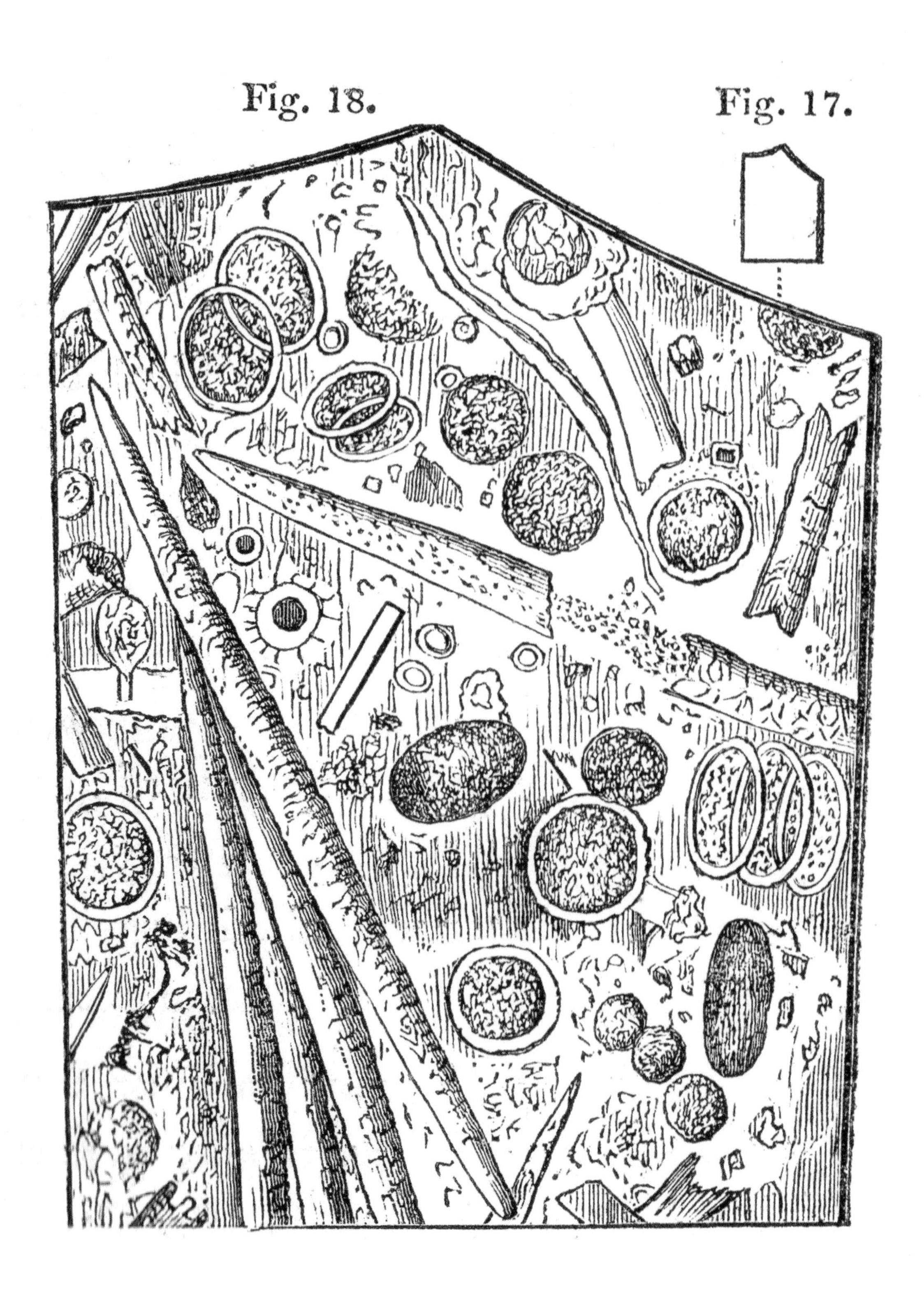
Fig. 18.
Fig. 17.

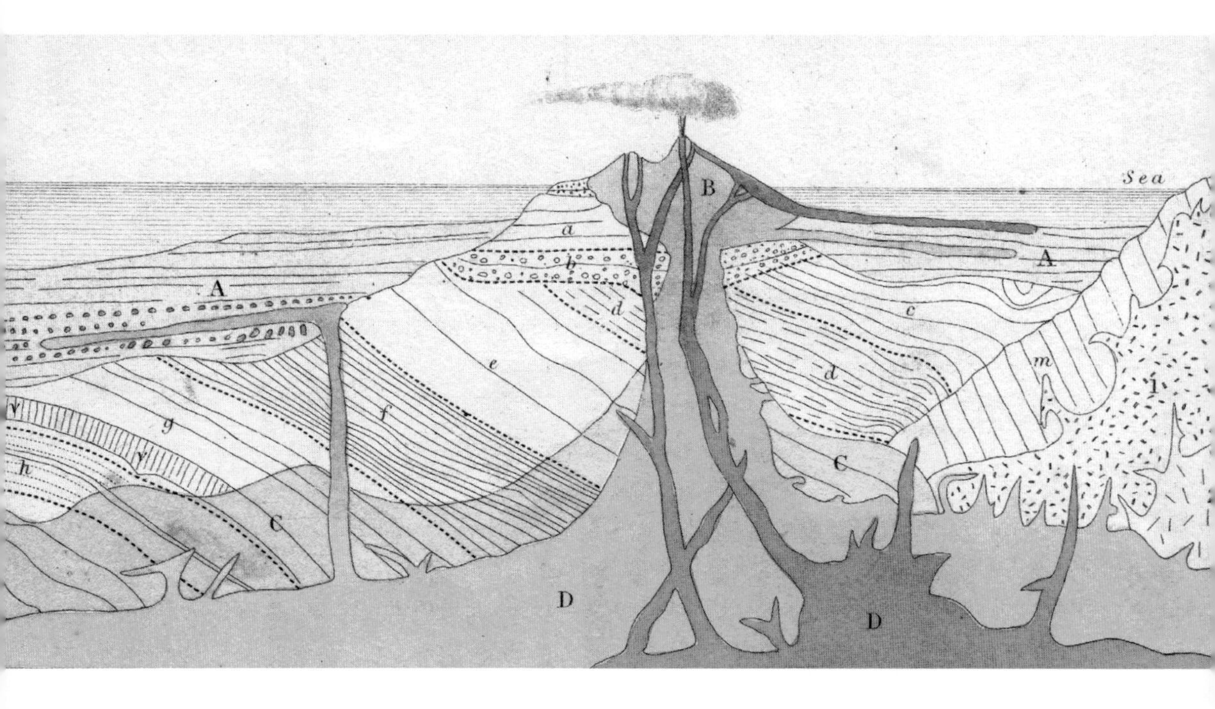


**Above: Charles Lyell (1797–1875), *Elements of Geology*, 1838; Section of Forfarshire, Scotland.**

Lyell was an influential geologist who extended the work of James Hutton (see page 15) in showing how geological processes have acted over extremely long periods to generate the landforms and even the climate we see today. His *Elements of Geology* was intended as a simplified, practical version of his multi-volume *Principles of Geology*, and it had a profound effect on Darwin's thinking—after all, the evolution of complex organisms needs long stretches of time to occur. Later, Lyell was also to sagely advise Darwin and Wallace when it suddenly became clear in 1858 that both had independently formulated the theory we now know as natural selection.

**Opposite: Charles Lyell (1797–1875), *Elements of Geology*, 1838; Fragment of semi-opal from the great bed of Tripoli, Bilin.**

**Elizabeth Philpot (1780–1857), Ichthyosaur skull after preparation, in a letter to William Buckland, 1833.**

Elizabeth Philpot was a fossil collector and paleontologist who excavated in Lyme Regis for most of her life and who became friend and confidant of the younger Mary Anning—forming the nucleus of a group of female paleontologists later portrayed in the 2020 film *Ammonite*. It was Philpot who discovered that the defensive ink produced by some marine molluscs could be resurrected from their long-dead fossils and she used it to depict, appropriately enough, fossils.

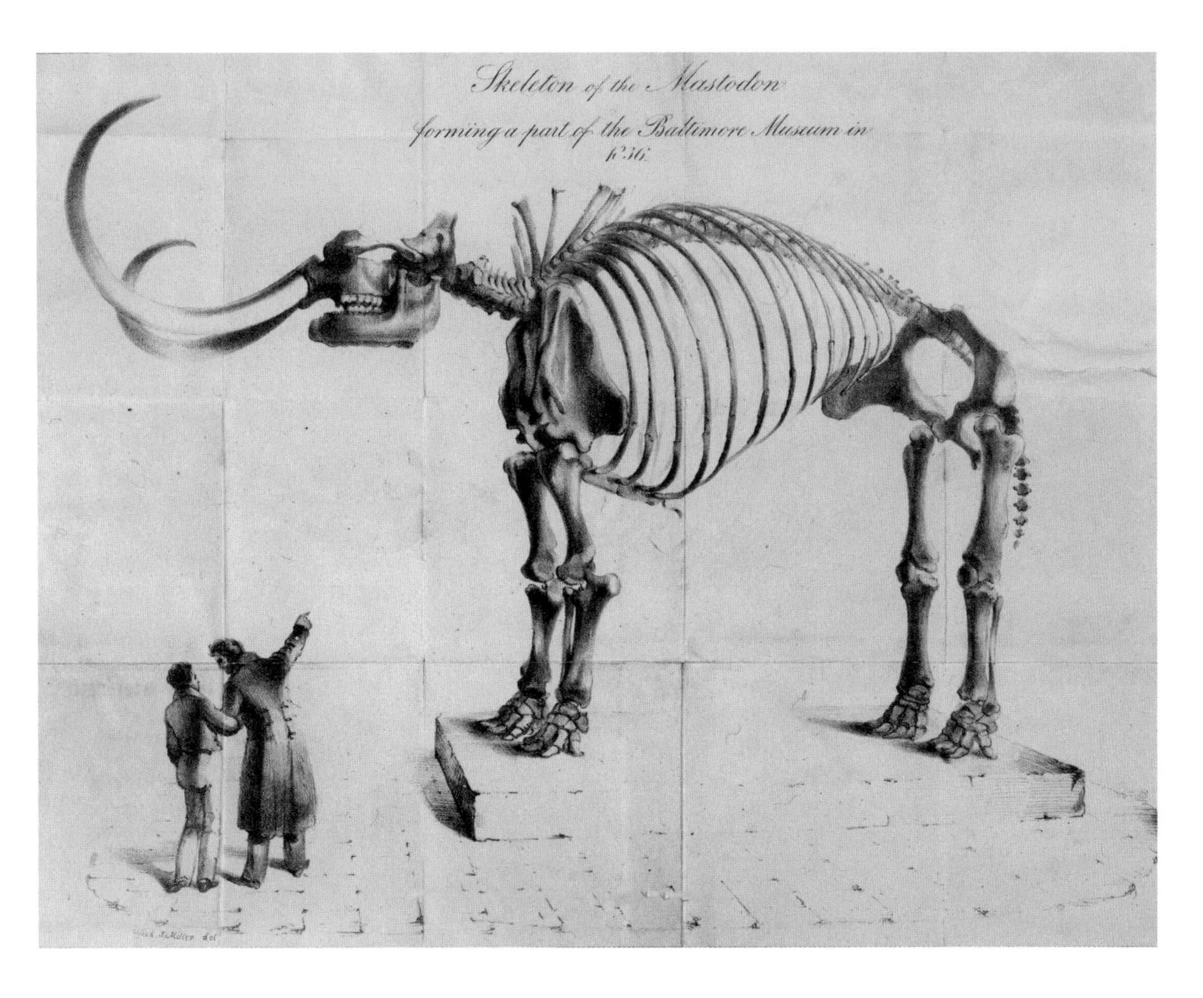


**Alfred Jacob Miller (1810–1874), *Skeleton of the Mastodon*, 1836.**

Spectacular ancient pachyderms were hauled from the muds of the Hudson and Ohio river valleys around the turn of the nineteenth century, including this specimen, whose unearthing was overseen by the Philadelphia naturalist and artist Charles Willson Peale. Many animals' scientific names, even the most familiar, have unexpected origins, none more so than this: George Cuvier, noticing the curved bumps in its teeth, named it Mastodon (breast-tooth).

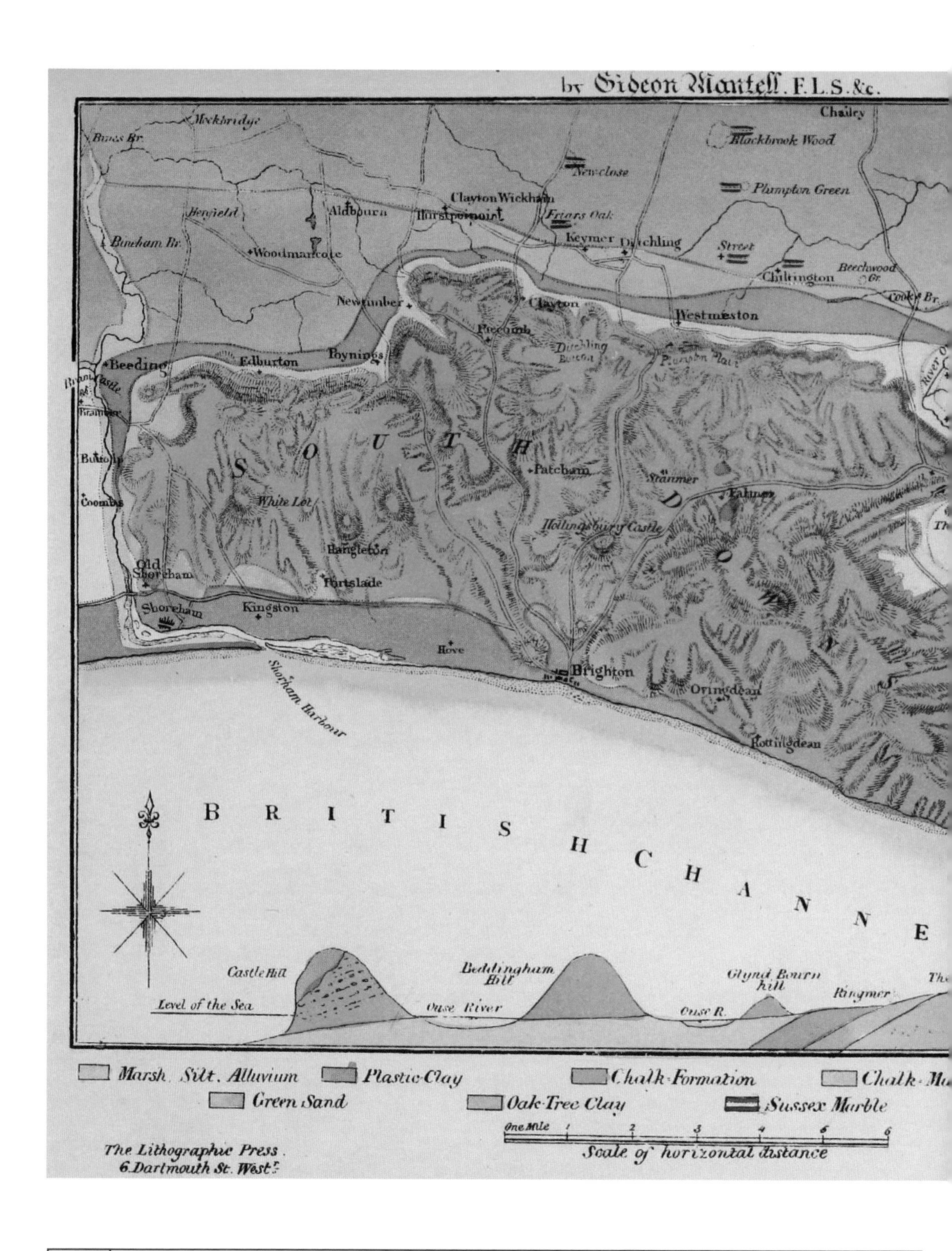
by Gideon Mantell. F.L.S. &c.
Chailey
Blackbrook Wood
Newclose
Plumpton Green
Clayton Wickham
Henfield
Aldbourn
Hurstpeirpoint
Friars Oak
Keymer
Ditchling
Street
Chiltington
Beechwood Gr.
Woodmancote
Newtimber
Clayton
Westmeston
Cook's Br.
Beeding
Edburton
Poynings
Ditchling Beacon
Plumpton Place
S O U T H D O W N S
Patcham
Stanmer
Falmer
White Lot
Hollingsbury Castle
Hangleton
Portslade
Old Shoreham
Shoreham
Kingston
Hove
Brighton
Ovingdean
Rottingdean
Shoreham Harbour
B R I T I S H C H A N N E
Castle Hill
Beddingham Hill
Glynd Bourn hill
Ringmer
Level of the Sea
Ouse River
Ouse R.
Marsh. Silt. Alluvium
Plastic Clay
Chalk Formation
Green Sand
Oak Tree Clay
Sussex Marble
One Mile 1 2 3 4 5 6
Scale of horizontal distance
The Lithographic Press.
6 Dartmouth St. West.

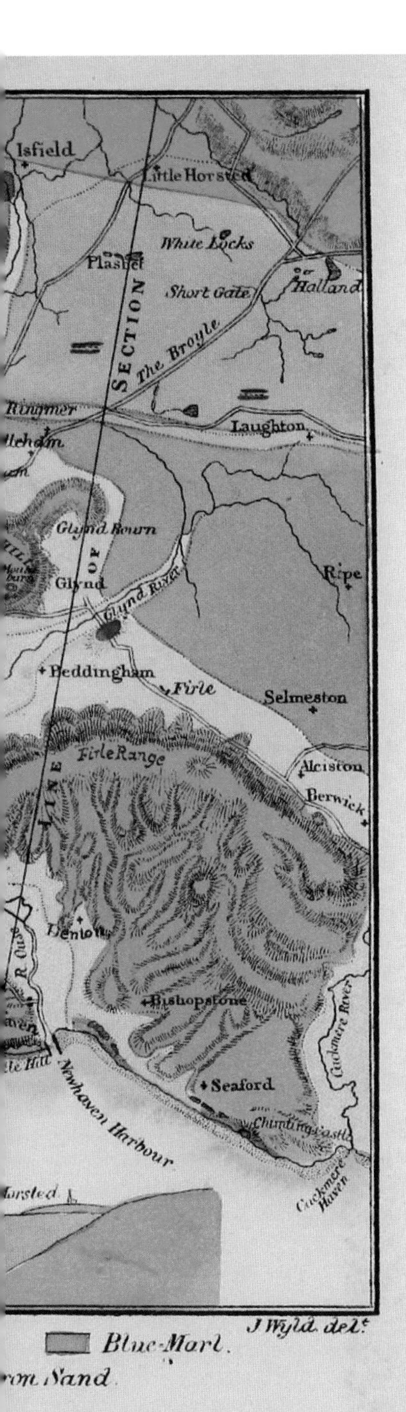


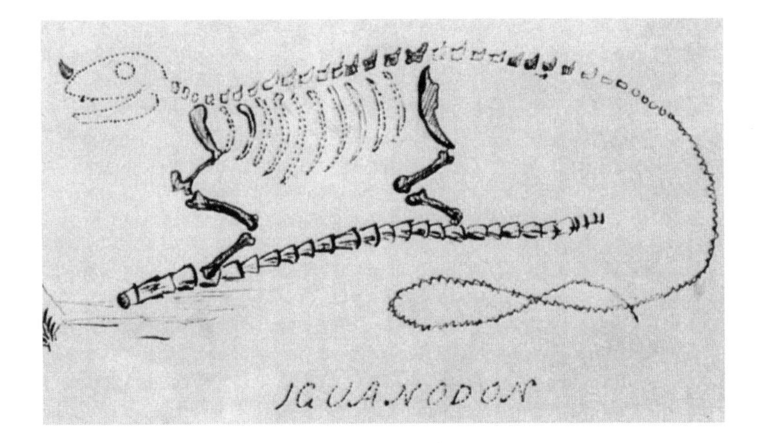


**Gideon Mantell (1790–1852), *The Fossils of the South Downs*, 1822; A Geological map of the South-Eastern part of Sussex (left); Reconstruction of an *Iguanodon* in pen and ink, 1834 (above).**

The story of the English doctor-cum-paleontologist Gideon Mantell is a rather sad one. Inspired by Mary Anning's discoveries, he is best known for the most complete characterization to date of a fossil dinosaur, *Iguanodon* (see page 26), and he also revealed much about the distant Cretaceous past of the landscape in which he lived. He showed that parts of his home county, Sussex, had at different times lain beneath the sea. His discoveries were often dismissed during his lifetime, even though some were supported by Charles Lyell and George Cuvier, and Richard Owen in particular was especially vitriolic in his attacks (see pages 26–27). Mantell's medical practice gradually failed, his wife divorced him—a rare thing at the time—he suffered a spinal injury in 1844, and eventually committed suicide by taking an overdose of opium. In a staggeringly mean-spirited show of disrespect, Owen arranged for Mantell's deformed spine to be excised *post mortem*, and displayed in London's Natural History Museum.

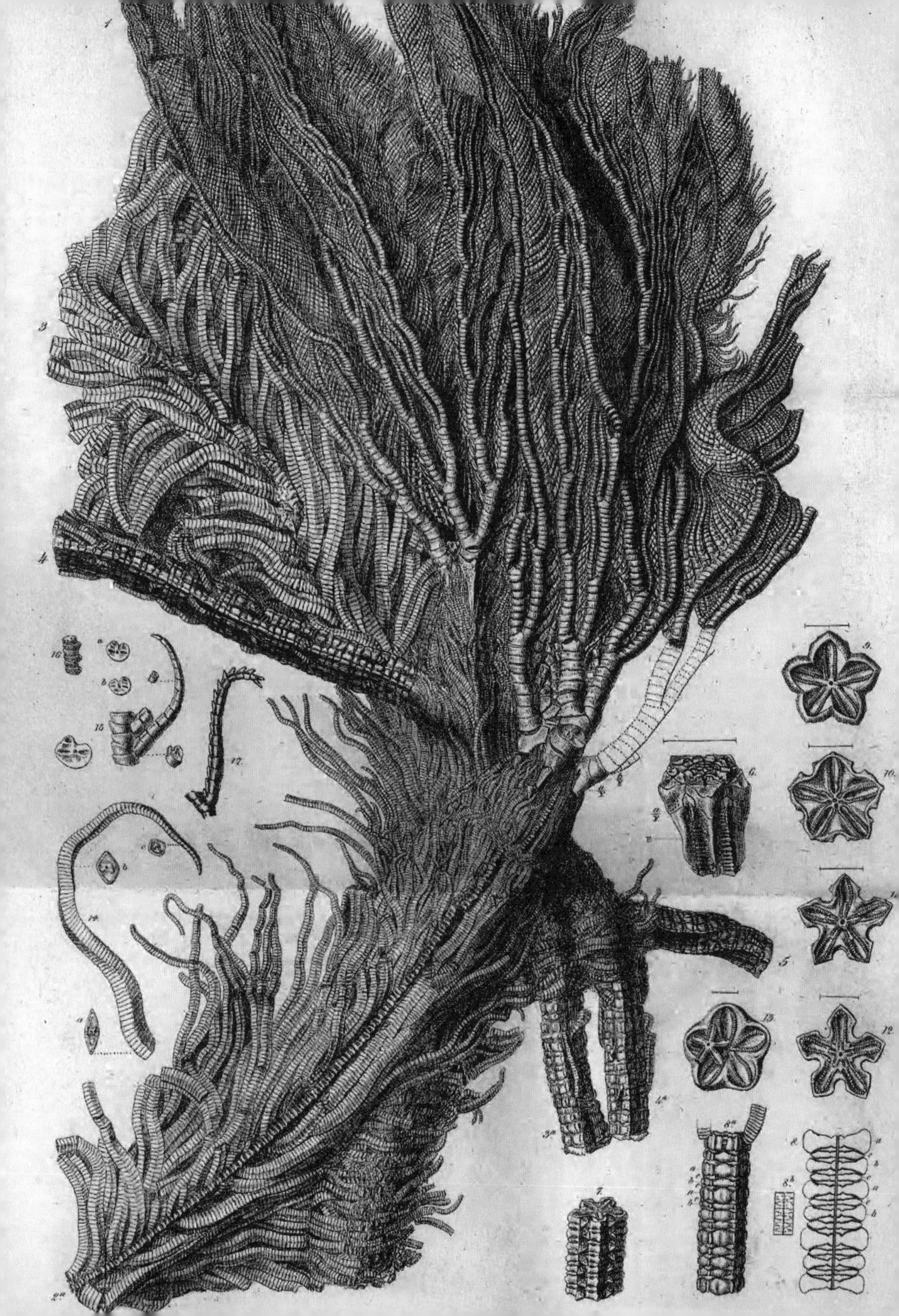

1
3
4
5
6
7
8
9
10
12
13
14
15
16
17

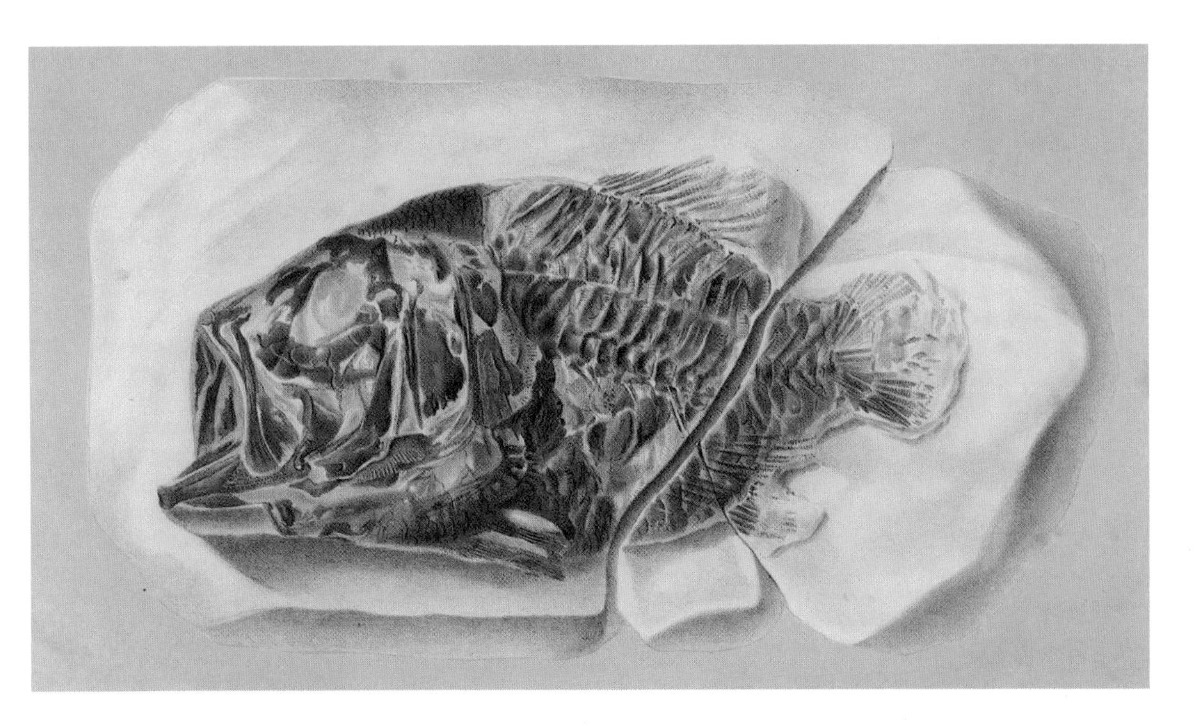

**Opposite: William Buckland (1784–1856), *Geology and Mineralogy Considered with Reference to Natural Theology*, 1836; Petrified crinoid (*Pentacrinites briareus*), from the Lias (Jurassic) at Lyme Regis, Dorset.**

Crinoids are marine invertebrates related to starfish and sea urchins, and they demonstrate the impressive longevity of some animal groups. Still widespread in today's oceans as "sea lilies" and their kin, crinoids have a fossil record stretching back almost 500 million years, and possibly longer. Although they were once more diverse and numerous than today, crinoids demonstrate that while the story of life on Earth may be headlined by huge, spectacular creatures and punctuated by cataclysmic extinctions, some humble yet beautiful animals have found evolutionary niches that allowed them to outlive many others.

**Above: Louis Agassiz (1807–1873), *Recherches sur les Poissons Fossiles*, vol. 4, 1835; *Beryx ornatus* fossil fish.**

Fish species currently outnumber those of all other vertebrate groups combined, and this has probably always been the case. As such, they provide a long and relatively continuous fossil record dating back over 500 million years. The Swiss-born paleontologist Louis Agassiz, working on fish especially between 1830 and 1845, was the first to study the jawless, armored ostracoderms (meaning "shell-skin") which are the most ancient fish known, thus doubling the length of the known history of backboned animals. This image, however, is of a much more recent (80–40 million years ago) jawed species.

# MARY ANNING (1799–1847)

## "The Greatest Fossilist the World Ever Knew"

*Mary Anning with her dog, Tray* **(oil on canvas by an unknown artist), circa 1842.**

Mary Anning was born and lived her life in Lyme Regis on the coast of Dorset, where the oblique stripe of Jurassic (201–145 million years old) rocks traversing England crumbles into the waves at its western terminus. Inspirational to many, distrusted by some, she was that rare thing: a nineteenth-century female scientist who received some credit for her work.

Mary's family was poor, although she was clearly extremely literate, being self-taught in biology and geology, and she learnt French expressly so she could read the works of George Cuvier (see pages 30–32). However, some belittled her as a working-class fossil hunter rather than a true scientist, because she sold her discoveries—it was her living after all—and rarely wrote about them herself.

Her first great excavation to catch the attention of the paleontological world was of the dolphin-shaped ancient reptile *Ichthyosaurus* (fish-lizard) when she was only twelve years old—one of the earliest specimens of an animal now obviously extinct—and in 1823

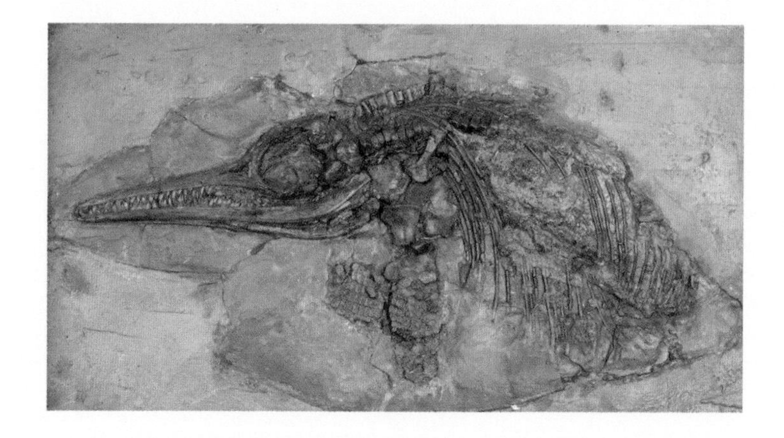

**Mary Anning (1799–1847), Partial skeleton of a young ichthyosaur, collected at Lyme Regis before 1836.**

**William Fox Talbot (1800–1877), "Geologists" at Chudleigh, Devon, February 1843.**

(It is thought that the man in the image may be Henry de la Bèche and the woman could be Mary Anning.)

she unearthed a more complete specimen. Her most spectacular fossil finds were perhaps those of *Plesiosaurus* (like a lizard) from 1823 onward, another ancient, large marine reptile, but in this case spindle-bodied, long-necked, and tiny-headed. These beasts were so strange that many could not believe they were not faked, including some scientists.

Not a "wife-assistant" like many of the under-credited, nineteenth-century female fossilists, Mary still often did not receive the credit she deserved and there remain many specimens, including sharks, bony fish, molluscs, and pterosaurs, which we suspect—though we cannot be certain—she discovered. She was an active paleontologist throughout her adolescent and adult life, and it seems more than one benefactor sold their collection to financially support her work. Tragically, Mary died at the age of forty-seven, but she had already lit the fuse of a scientific explosion whose echoes rumble down to the present.

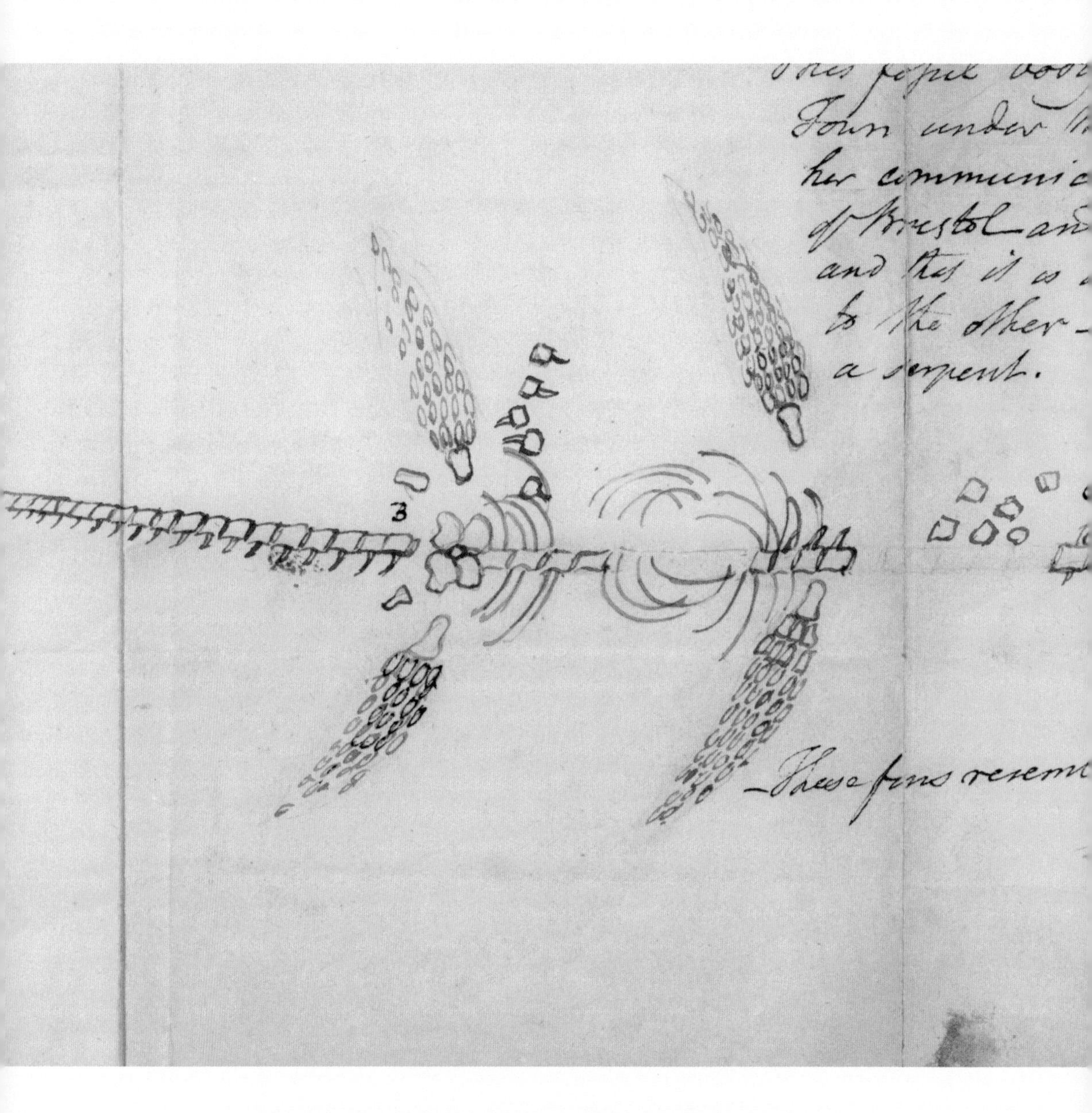


**Mary Anning (1799–1847), First sketch of the *Plesiosaurus*, 1824, discovered by Anning on the Jurassic Coast, Dorset, in 1811.**

ack vein Cliff. in Decr 1823. or Jany 1824. and was by

by a rude drawing of which this is a copy to Mr Johnson

rs. in her ~~letter~~ she states that it is very well preserved

8 or 9 feet long. and 4 feet from the point of one f

head only 6 or 7 Inches — perhaps less. — and resembles

D

re of the fossil called Plesiosaurus.

SKELETON OF TRACHODON (HADROSAURUS) AS RESTORED BY B. WATERHOUSE HAWKINS.

**Opposite: Benjamin Waterhouse Hawkins (1807–1894), Skeletal reconstruction of *Trachodon* (now *Hadrosaurus*), 1868.**

Discovered in 1838 in Haddonfield, New Jersey, this almost complete skeleton was reconstructed three decades later to become the first dinosaur skeleton fully mounted for display at the Philadelphia Academy of Natural Sciences. The reconstruction was led by naturalist and sculptor Benjamin Waterhouse Hawkins, the designer of the Crystal Palace Dinosaurs in London (see page 64), who also molded a plaster skull based on those of closely related species, to sit atop its neck. As the first publicly displayed "full" dinosaur skeleton, *Hadrosaurus* gained fame in the United States, and in 1994 was made the state fossil of New Jersey.

**Above: Edward Newman (1801–1876), "Note on the Pterodactyle Tribe Considered as Marsupial Bats," *The Zoologist*, vol. 1, 1843; *Pterodactylus crassirostris* (above) and *Pterodactylus brevirostris* (below).**

Edward Newman was an English writer and entomologist, who briefly and famously diverted his attention to the biology of pterosaurs. There had been claims that some pterosaur fossils show evidence of fur, so Newman authored a paper, not entirely seriously, proposing that pterosaurs were, in fact, marsupial bats. Although this theory never gained much support, it did somewhat prefigure arguments over a century later about the warm-bloodedness of dinosaurs and pterosaurs.

**Richard Owen (1804–1892), *Odontography*, vol. 2, 1840; Labyrinthodont tooth.**

For all his faults, Richard Owen was the preeminent comparative anatomist of the mid-nineteenth century. As Conservator of the Hunterian Museum of the Royal College of Surgeons and effectively the founder of London's Natural History Museum, Owen's influence dominated paleontology for decades—and he published a large corpus of work. This image is a cross section of a tooth from a "labyrinthodont" amphibian, and clearly explains the name conferred upon that group. The labyrinthodonts were the dominant early amphibian tetrapods ("four-footed" land vertebrates) which prospered from 400–150 million years ago.

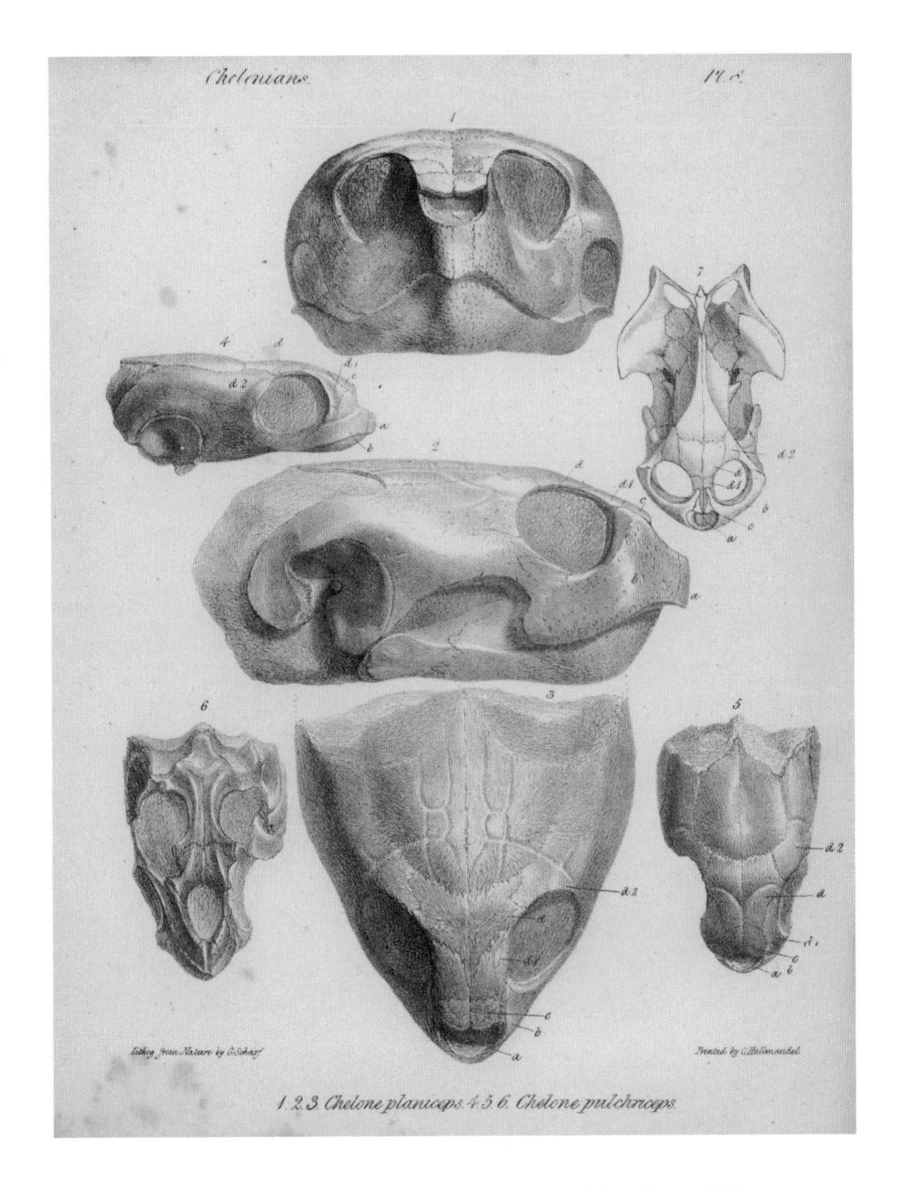


**Richard Owen (1804–1892), *A History of British Fossil Reptiles*, vol. II, 1849–1884; *Chelone planiceps*.**

The chelonians, turtles and tortoises, present scientific challenges for both morphologists and evolutionary biologists. They are, of course, structurally unusual—their ribs form the inner layer of their carapace, so that their pelvis, shoulder blade, and associated bones lie *inside* their rib cages, unlike other vertebrates. Also, they lack the windows in the side of the skull often used to classify land vertebrates, leading to a classificatory conundrum which waited until the twenty-first century for its resolution (as shown on page 234).

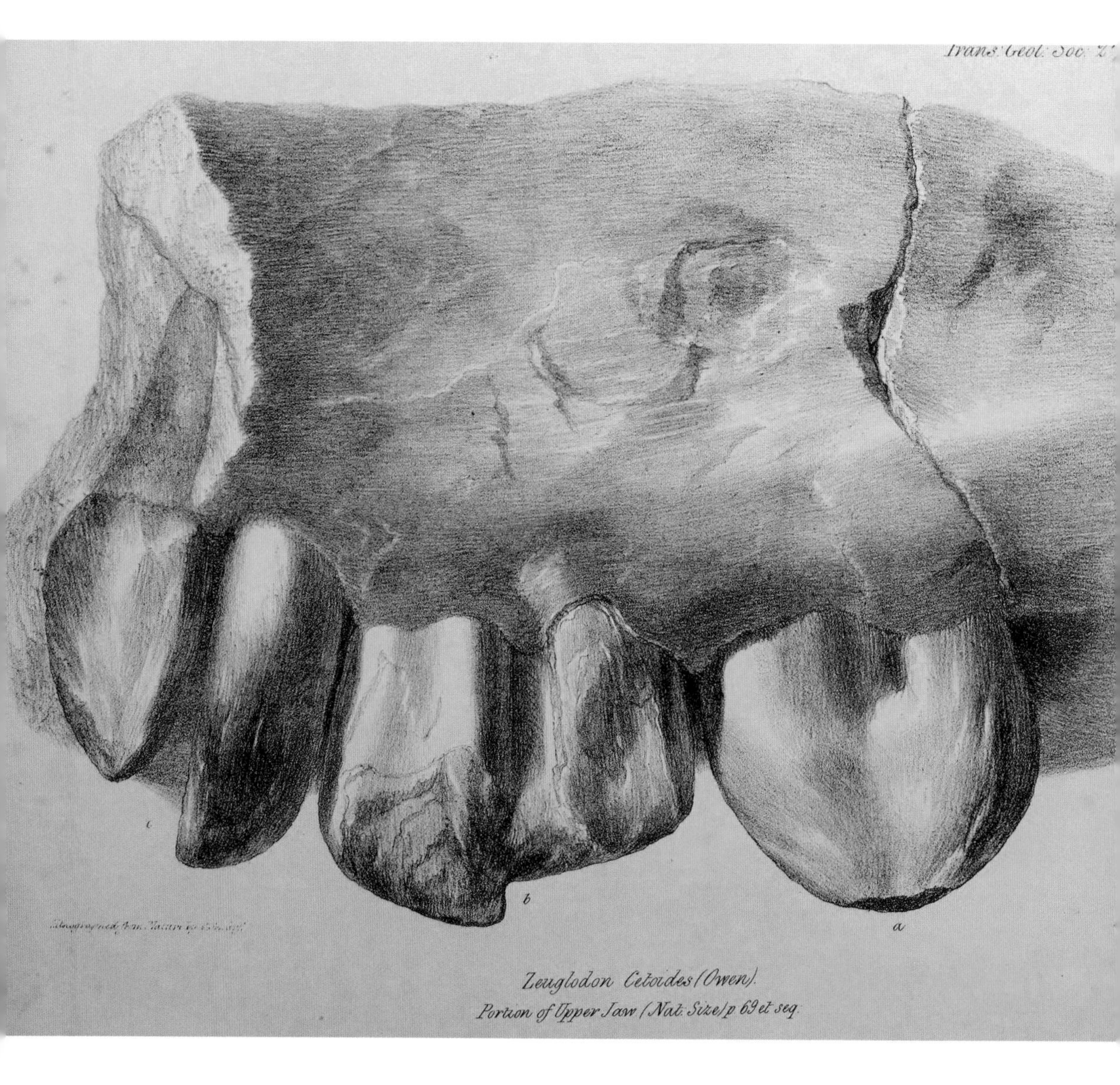


**Richard Owen (1804–1892), Fragment of upper jaw of *Zeuglodon cetoides* (now *Basilosaurus*), *Transactions of the Geological Society of Pennsylvania*, vol. 6, 1842.**

Once the name of a species has been agreed, there are strict rules about whether it can subsequently be changed, which explains why this ancient whale is to this day called "king-lizard" or *Basilosaurus* (see page 26). Richard Owen was one of the anatomists who pointed out that this species has teeth with multiple roots, a feature characteristic of mammals. We now know *Basilosaurus* lived 40–35 million years ago, some time after the extinction of the giant marine reptiles.

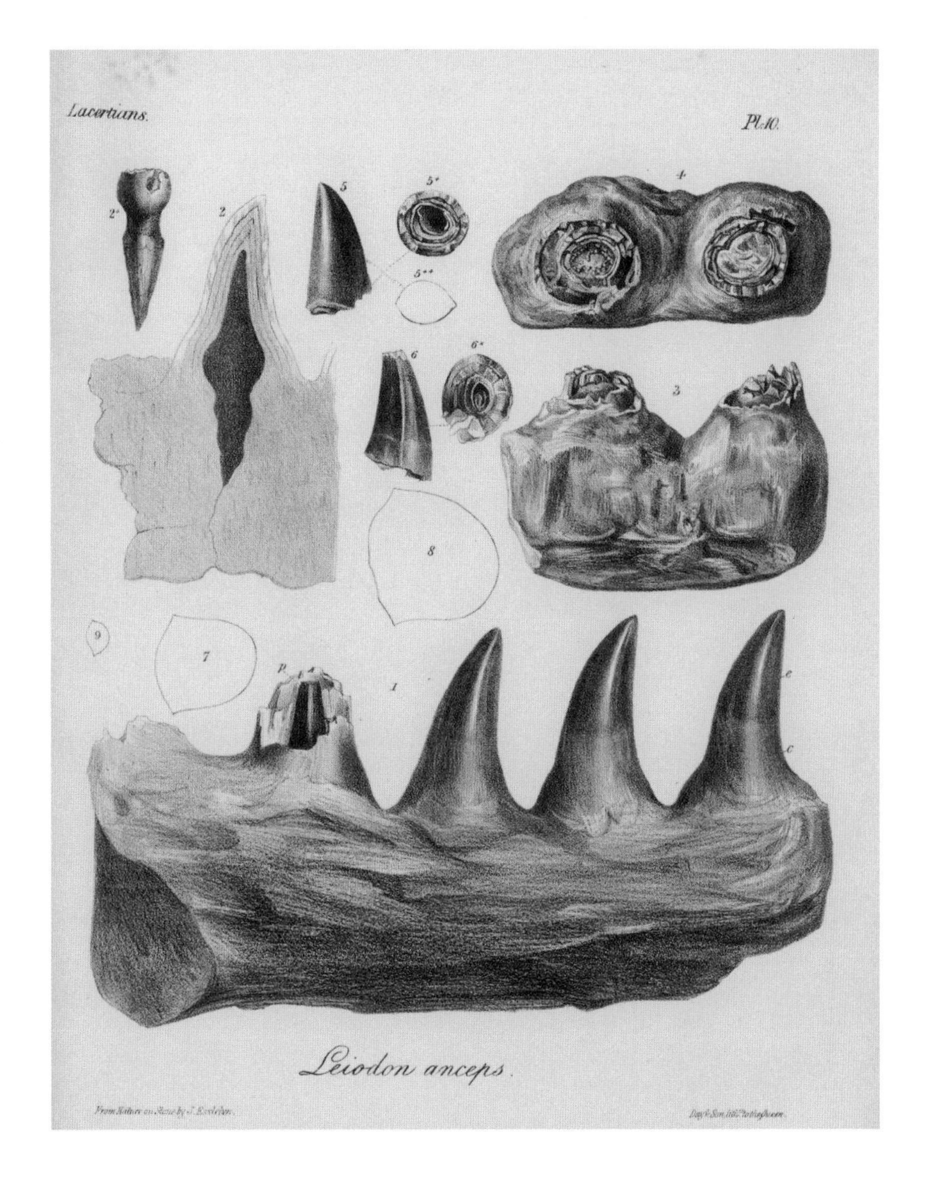


**Richard Owen (1804–1892),** ***A History of British Fossil Reptiles*****, vol. II, 1849–1884;** ***Liodon anceps*****.**

*Liodon* remains something of a mystery today, and is still known only from teeth and fragments of jaw. Unlike mammals, it has simple, single-rooted teeth, and is thought to be a species of mosasaur, a large marine reptile probably related to lizards and snakes, like the "Beast of Maastricht" (see page 32).

**Sectioned ammonite, from Madagascar (private collection).**

Ammonites are molluscs closely related to modern cuttlefish, squid, and octopuses, which flourished in the oceans from approximately 400 million years ago until the late-Cretaceous extinction event 66 million years ago. They were especially prevalent in the Jurassic Period and fossils of their spiral shells, with their characteristic internal "septa" or dividing partitions, were the mainstay of fossil hunters on the Dorset coast, including Mary Anning (see page 52). Ammonites were a diverse and abundant group, which underwent rapid evolutionary change, making them ideal for estimating the date of the rock strata in which they are fossilized.

**Top: Adam Clark Vroman (1856–1916), *The Big Log Cannon in the Petrified Forest*, photographed in 1895. Above: the Petrified Forest of Arizona, photographed in 2012 by Natasha de Vere & Col Ford.**

The Petrified Forest of Arizona was first brought to the attention of scientists in 1851 when a series of government-led expeditions explored this remote, dry region. The fossils in what is now the Petrified Forest National Park were deposited in the Triassic Period around 220 million years ago, a time when today's continents were fused into a single gigantic supercontinent, Pangea. The trees were washed down toward the western shores of Pangea where a minority of them became fossilized, some with such a high level of preservation that individual cells may be discerned under a microscope.

Left: Benjamin Waterhouse Hawkins (1807–1894), Invitation card for the dinner at the Crystal Palace *Iguanodon* Mold in 1853.

Below: The dinner held by Benjamin Waterhouse Hawkins in the *Iguanodon* Mold in 1853 (shown in an unattributed woodcut from 1854 that appeared in the *Illustrated London News*). Richard Owen is at the head and Hawkins in the center.

**Cleaning the Crystal Palace Ichthyosaur, 1930.**

An unexpected gem waiting to be discovered in Crystal Palace Park in south London, these cement-and-brick monsters are unique—a veritable crystallization of mid-Victorian paleontological knowledge. They include representations of giant Mesozoic land, marine, and flying reptiles, which correspond only occasionally to modern conceptions of these animals' likely appearance, although their artistic ambition cannot be questioned. One was even large enough for a dinner party to be hosted within its mold (see opposite). The shimmering Crystal Palace built for the 1851 Great Exhibition is now long gone, and Benjamin Waterhouse Hawkins' plans for an American equivalent of the dinosaurs never came to fruition, but these most unusual English suburban residents survive, albeit with almost continual renovation and repair.

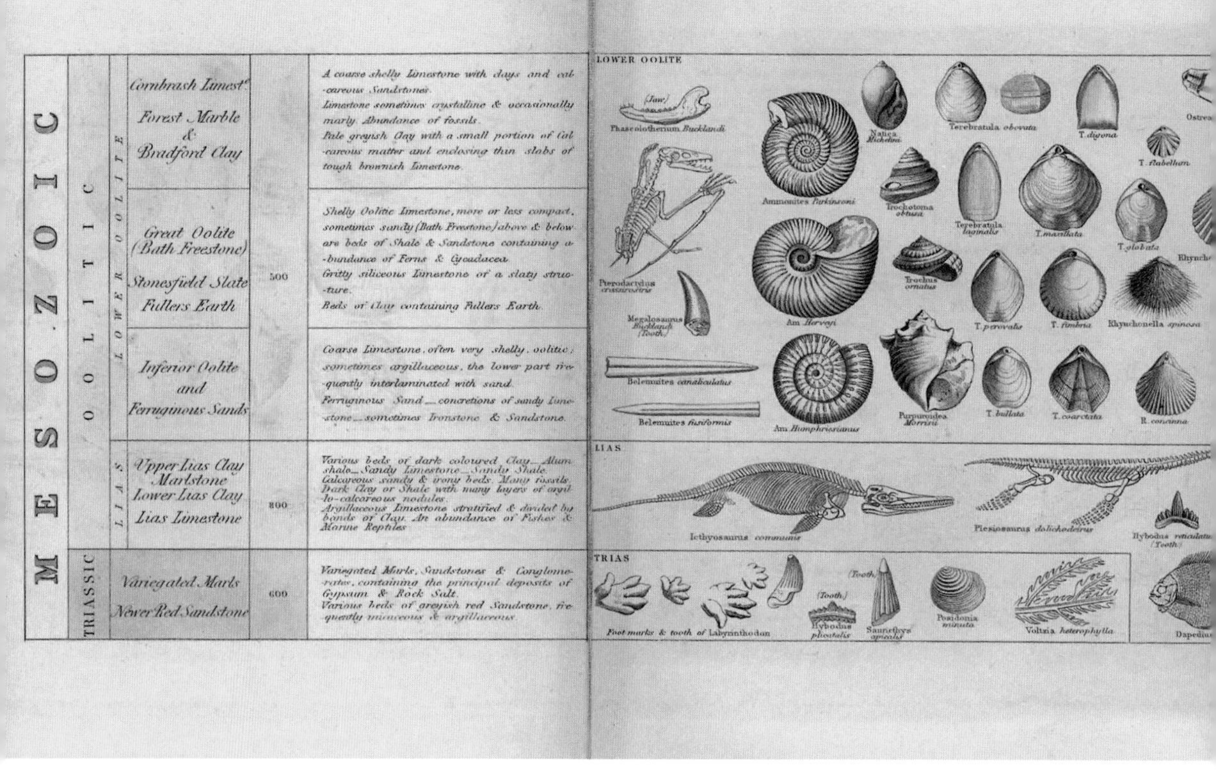


**Above: Joseph Wilson Lowry (1803–1879), *Tabular View of Characteristic British Fossils, Stratigraphically Arranged*, 1853; Mesozoic British fossils.**

Montages such as this one (intriguingly published by the Society for Promoting Christian Knowledge) fixed in the public mind William Smith's discovery that each layered rock stratum contains distinctive, characteristic fossilized fauna (see page 34). The image also correctly implies that the fossil record is dominated not by spectacular reptiles, but by more common, readily fossilizable, humble molluscs.

**Opposite: Richard Owen (1804–1892), *Geology and Inhabitants of the Ancient World*, 1854; *Megalosaurus*.**

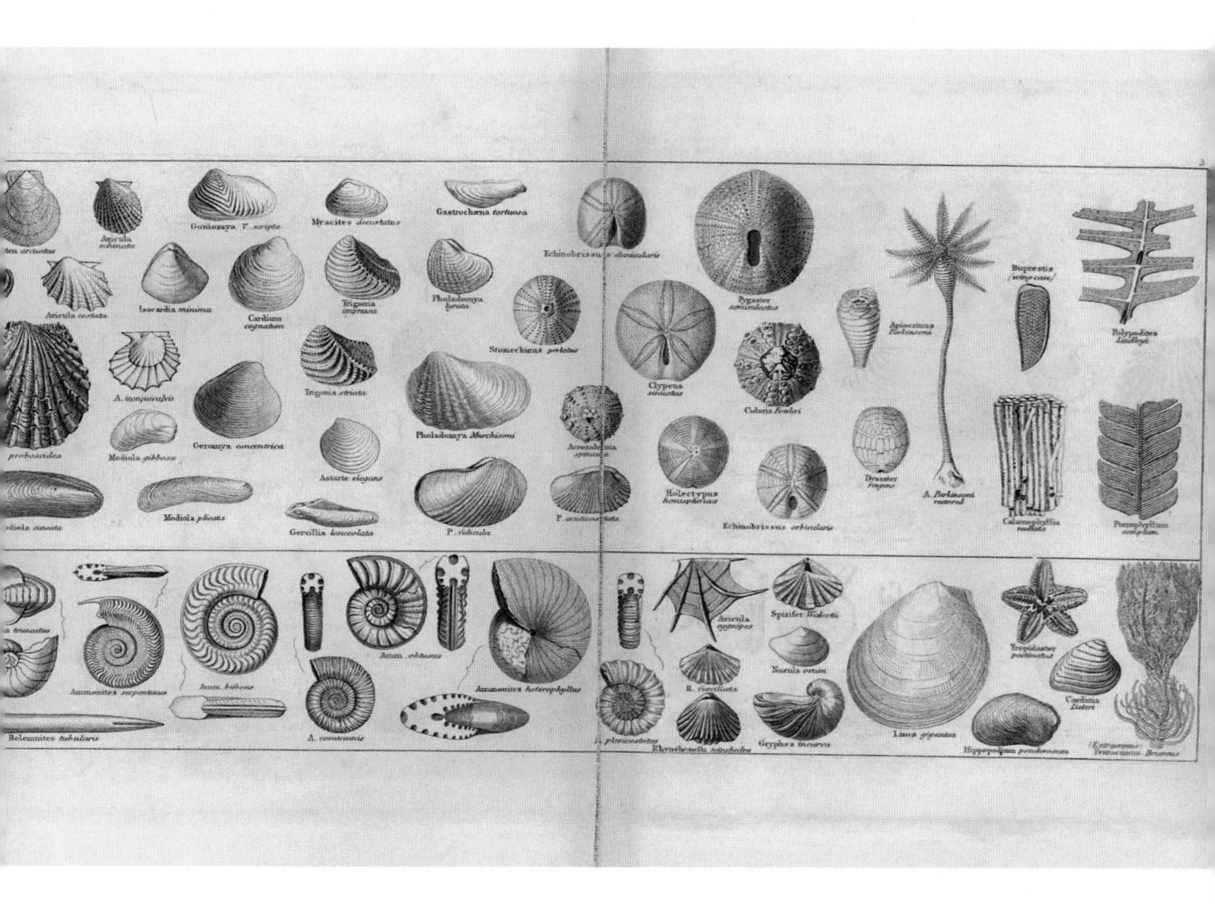

No. 7. Megalosaurus.

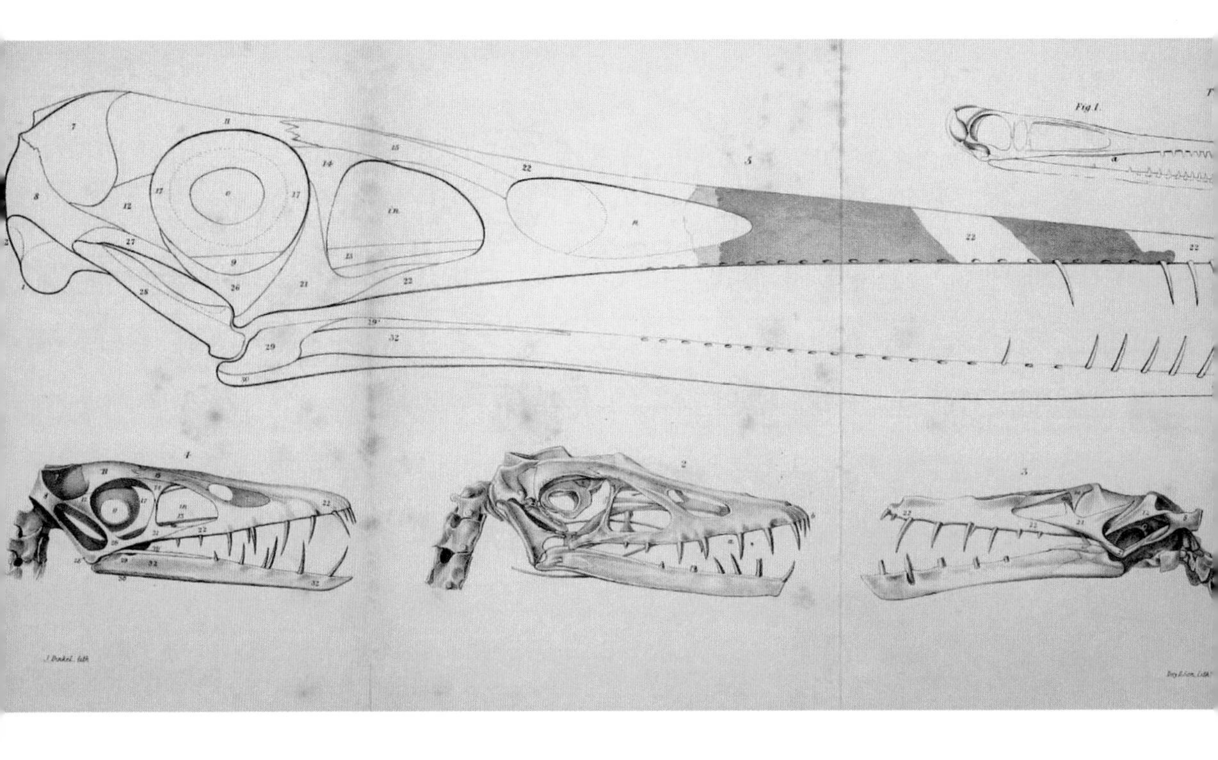

**Richard Owen (1804–1892),** ***A Monograph on the Fossil Reptilia of the Cretaceous Formations*****, 1851–64; Skulls of** ***Pterodactylus longirostris, crassirostris*****, and** ***compressirostris*****.**

The enigmatic genus *Pterodactylus* or "wing-finger" (see page 41) encompassed a variety of winged forms, all probably small-scale carnivores. They are now recognized as representing just a subset of a more diverse, larger group, the pterosaurs (wing-lizards) whose striking morphology remains a challenge to today's paleontologists as they attempt to explain how they could fly, feed, and breed without shattering their spindly frames (as shown on page 174).

**Dr Paul Taylor (b.1952), from Giles Miller, *Conodonts: the Most Controversial Microfossils?*, 2013; Scanning electron microscope image of conodonts from the Silurian Period of Gotland, Sweden.**

The conodonts (cone-teeth) were, and still are to some extent, one of the greatest enigmas in paleontology. Readily extracted from marine deposits as old as 485 million years, these tiny cones and comb-shaped spicules have defied classification ever since they were first claimed to be fossilized fish teeth in 1856. They have been assigned to almost every animal group since, but in 1980 the fossil of an animal bearing conodont-like denticles was finally discovered. Aspects of its structure suggest it was a vertebrate, or at least a close relative of vertebrates, although it is not even certain which way up the animal swam. The conodonts survived until the Triassic-Jurassic extinction event 201 million years ago, in which the sea became catastrophically acidified by carbon dioxide spewing from volcanoes. It is notable that the current man-made increase in atmospheric carbon dioxide is also causing marine acidification, and this could prove to be more disruptive to life on Earth than global warming.

# CHARLES DARWIN (1809–1882)

## Fossil Hunting in South America

Less famed than his later visit to the Galapagos Islands, Charles Darwin's fossil-hunting escapades on the South American mainland were just as important in his later formulation of the theory of natural selection.

The voyage of H.M.S. *Beagle* lasted from 1831 to 1836, and the seasickness-prone Darwin collected his South American fossils ashore between 1832 and 1834, on both the east and west coasts of the continent. Some specimens Darwin discovered himself, while others he bought from locals, including the most striking, *Toxodon platensis*, an elephant-sized hoofed mammal with protuberant, rodent-like gnawing teeth. He also unearthed novel specimens of fossil horses, giant sloths, and *Glyptodon*: heavily armored relatives of modern-day armadillos.

Darwin commented that he was "much struck" by the similarities and differences between the living and extinct inhabitants of South America, and wondered how the Earth was repopulated, apparently with new

**Skull of *Toxodon platensis*, purchased by Charles Darwin in Uruguay.**

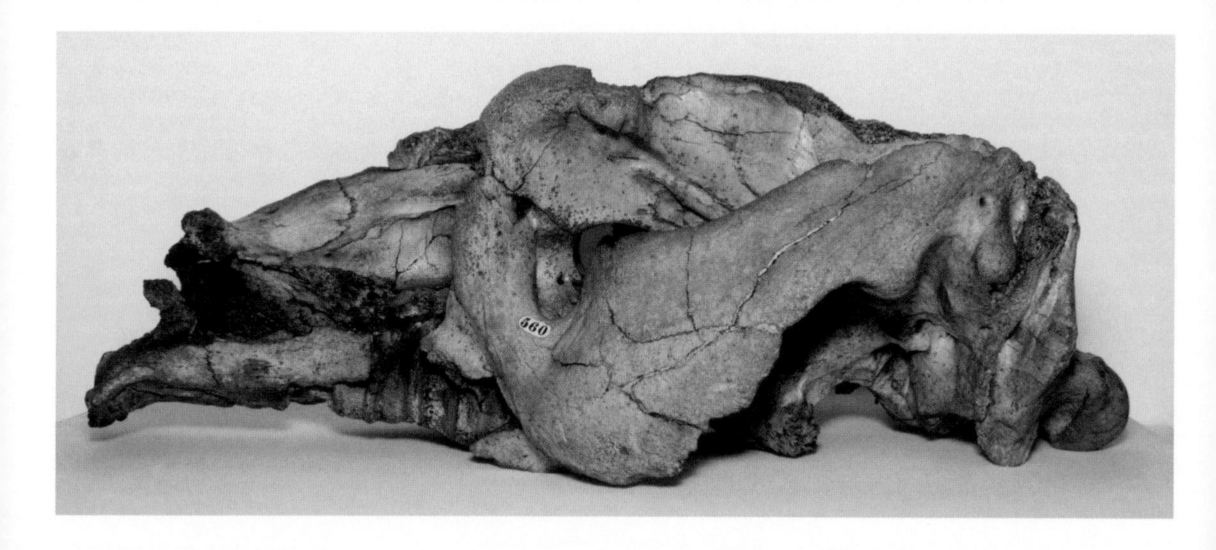

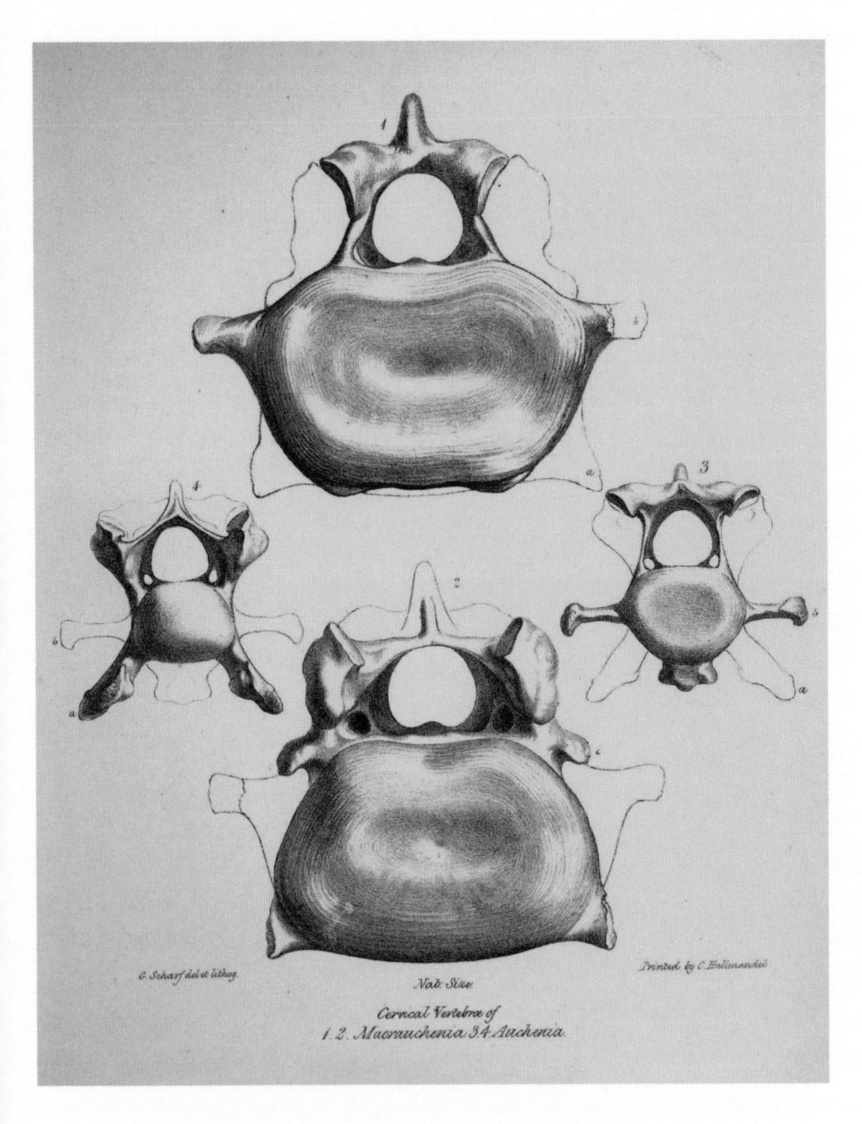


**Richard Owen (1804–1892), from Charles Darwin, *The Zoology of the Voyage of H.M.S. Beagle*, Part I, 1840; Cervical vertebrae 1 of *Macrauchenia patachonica* and *Auchenia llama*.**

forms, when species had become extinct. Also, like many before him, he was intrigued to find marine fossils, corals, and shells, high in the mountains—in this case the Andes.

The fossils Darwin collected did not reach Europe on the *Beagle* but were safely dispatched back to the Royal College of Surgeons in London, eventually to be analyzed by Richard Owen. Although the young Darwin had misidentified many of the specimens, Owen was struck by their scientific importance. Unfortunately, many were later destroyed in a bombing raid on London during the Second World War.

**Top: Fossil coelacanth, found at Wattendorf Plattenkalk, in Franconia, and dating from the late Jurassic Period. Above: Modern coelacanth, *Latimeria chalumnae*, caught off the Comoro Islands, near Madagascar, in the mid-1960s.**

The coelacanth is the most famous "living fossil." Fossil coelacanths, first discovered in the nineteenth century and now known to date from as early as 400 million years ago, were for many decades thought to be all humans would ever see of this creature, until a specimen was fished from the Indian Ocean in 1938. Coelacanths are now known to inhabit deep-sea habitats across the Indian Ocean and Indonesia, and while a few dazed individuals have been hauled live to the shallows, their behavior in their natural habitat has now been filmed by submersibles. Coelacanths are scientifically important because they are one of the few "sarcopterygian" (lobe-finned) fishes in existence, and along with lungfish represent the closest living relatives of land vertebrates.

**Fossil of cynodont *Galesaurus planiceps*.**

Mammalian evolution has a long separate pedigree from that of modern reptiles and birds, and a distinct lineage called the "synapsids" is evident from approximately 310 million years ago. From that time a series of successful groups sequentially acquired the characteristics we now think of as typically mammalian. Behind their eye socket lies a single large bony window, the "synapsid" hole, which distinguishes them from most reptiles which have two holes, and turtles and tortoises which have none. This hole developed for the attachment of the powerful jaw muscles which are one of the features that have made mammals so successful. The mammalian ability to chew means food can be divided into small fragments which are digested more rapidly and efficiently. Some have even suggested that cynodont skeletons show evidence of whiskers, also raising the possibility that these beasts were furry.

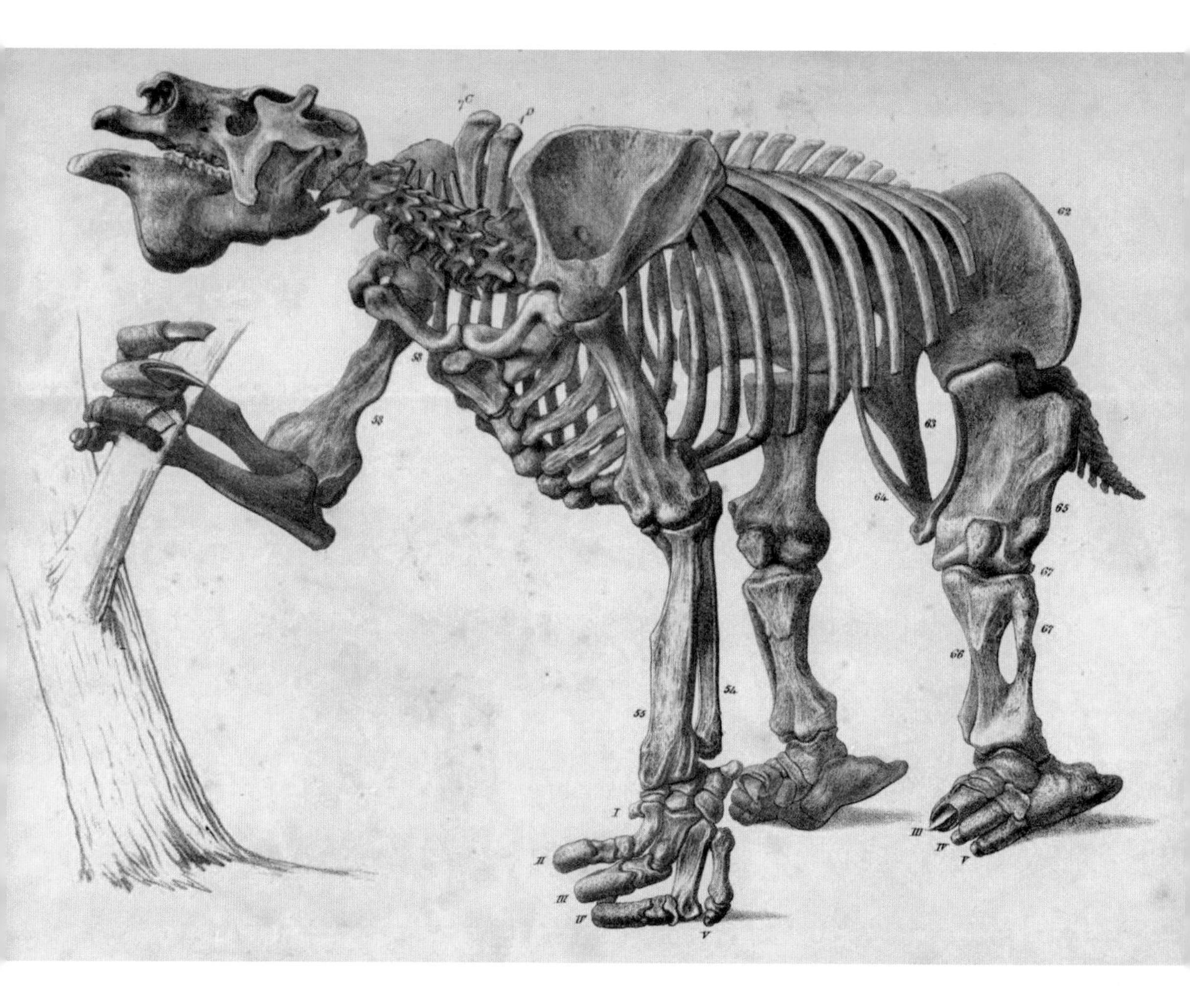

**Joseph Dinkel (1806–1891), from Richard Owen, *Memoir on the Megatherium, or Giant Ground-Sloth of America*, 1861.**

Extinct South American giant sloths were a favorite of nineteenth-century paleontologists, and this image is by the Austrian scientific artist Joseph Dinkel. *Megatherium* (huge-beast) was one of the largest land mammals ever, similar in weight to an African elephant, although fossil trackways suggest that it was, remarkably, capable of bipedal locomotion. *Megatherium* became extinct approximately 10,000 (0.01 million) years ago.

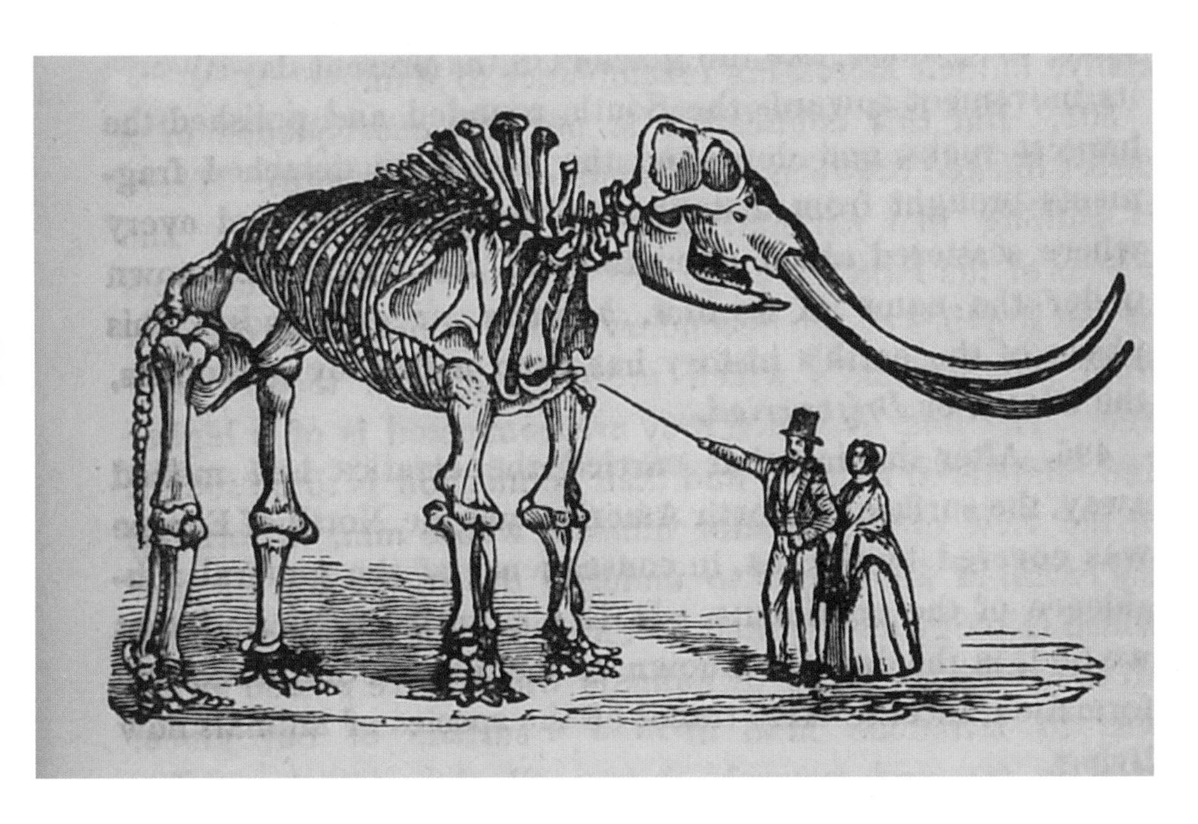

**Louis Agassiz (1807–1873) and Augustus Addison Gould (1805–1866),** ***Principles of Zoology: Touching the Structure, Development, Distribution, and Natural Arrangement of the Races of Animals*****, 1848; Mastodon skeleton.**

Probably the only land mammals heavier than elephants, *Megatherium* were a few species of mammoth and mastodon. Part paleontology, part social history, this image depicts the typical Victorian couple on their learned sojourn to a museum of natural history.

**The fossil snake *Archaeophis proavus*, from Monte Bolca, Verona, Italy.**

Although they are generally accepted to have evolved from lizards, much remains controversial about the origin of snakes. Many scientists have argued that their limblessness and extreme body shape indicate a past phase of adaptation to an unusual environment, possibly underground or aquatic. However, other lineages of lizards and amphibians have evolved a similar body plan, albeit without snakes' bizarre skull morphology, and this suggests that no such unusual past history is necessary to explain it. Described in 1859, this snake fossil dates from the Eocene Period, 56–34 million years ago—relatively late in the group's evolutionary history.

**Charles Lyell (1797–1875), *Principles of Geology*, vol.1, 1872; Tower of the buried church of Eccles, Norfolk, shown in 1839 (top) and 1862 (above).**

Charles Lyell was keen to demonstrate that, although the processes driving change in the Earth remain the same, there is evidence of that change all around us and it can sometimes be discerned over a human timeframe. In these images he demonstrates the transience of some landforms, as erosion exposes a church tower over the course of two decades. Since these images were created, this same church tower has been lost to the sea by coastal erosion.

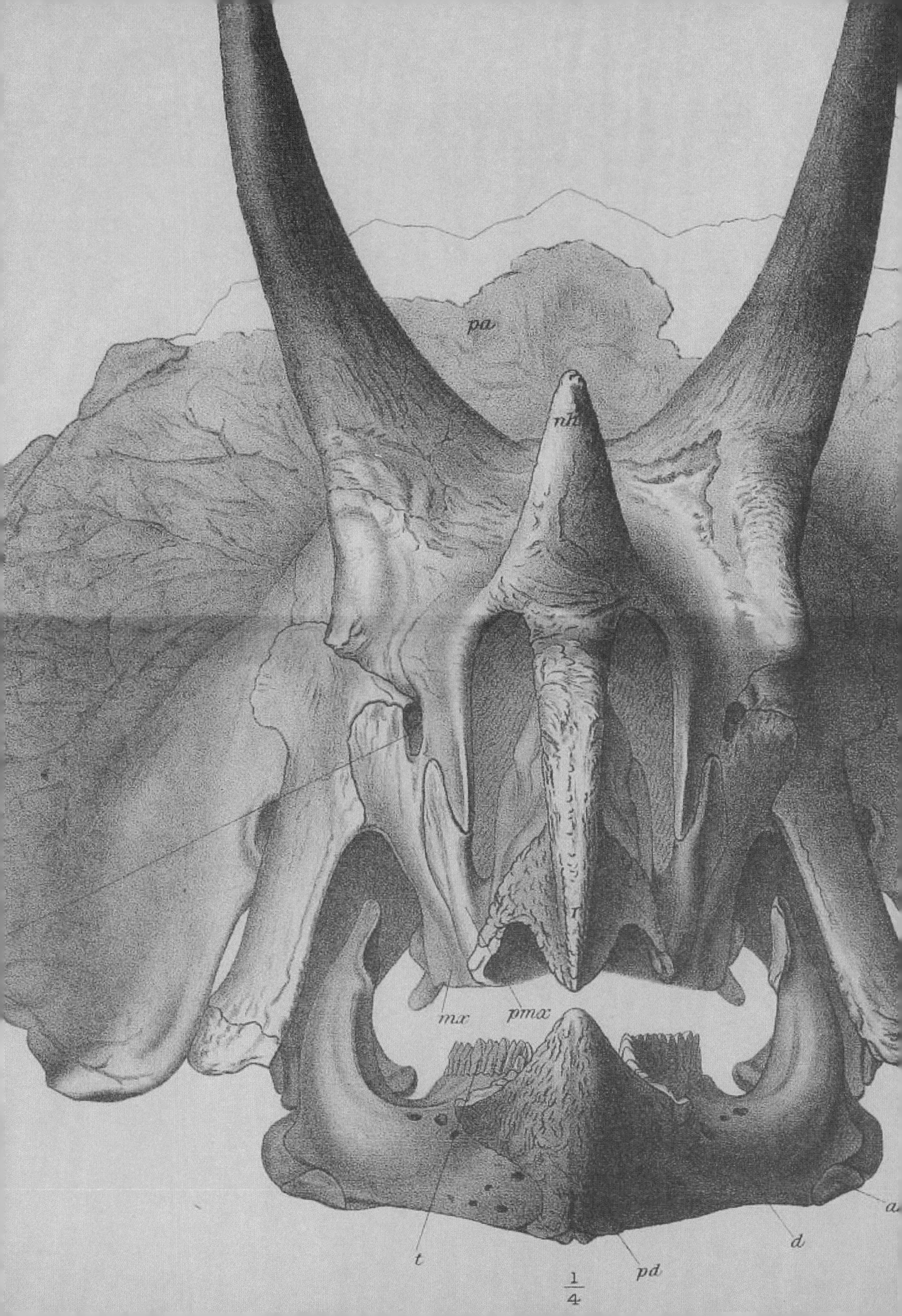
pa
n
mx
pmx
a
d
t
pd
1/4

Chapter Two

# A PRETERNATURAL SELECTION

1860–1920

John Bell Hatcher (1861–1904), *The Ceratopsia*, 1907;
Skull of *Triceratops prorsus*.

# A PRETERNATURAL SELECTION

## 1860–1920

The fossils of the proto-bird *Archaeopteryx* are almost too good to be true. And by sheer coincidence, this most striking "missing link" fossil—a previously unknown intermediate stage between known species—was discovered in 1860, the year after the publication of Darwin's *On the Origin of Species*. Many scientists realized evolution implied the fossil record should contain "missing link" fossils, but no one expected one so soon.

A description of this famous little creature, *Archaeopteryx lithographica* or *Urvogel* (meaning "ancient wing written in stone" or "first bird"), by German paleontologist Hermann von Meyer (1801–1869) followed one year later. Initially, the fossil imprint of only a single feather was found (see page 86), but this was followed by a partial skeleton the next year, and the almost-complete "Berlin specimen" in 1874—all etched with amazing fidelity into the finely layered limestone of the quarries near Solnhofen, in Bavaria, whence, in fact, all twelve known fossils of this species come. *Archaeopteryx* had toothy jaws rather than a beak, a long tail rather than a short one, claws

**Chromolithograph of Solnhofen stone quarry, in Bavaria, Southern Germany, 1889.**

on its wings, and a pelvis intermediate between those of a dinosaur and a bird—yet there it lies, frozen in stone, radiating a halo of unmistakable feathers, the asymmetrical feathers typical of birds that *fly*.

Still, for many scientists further evidence was needed that there had been enough *time* for the evolution of all these fossil species to take place, and a major scientific argument started in 1862 when the British physicist William Thomson (or Lord Kelvin, 1824–1907) first estimated the age of the Earth. Kelvin was an expert on heat—the "Kelvin" is now the international scientific unit of temperature—and he calculated how long it would take an Earth-sized ball of molten rock (starting at 7,000°F, just below 4,000°C) to cool down to its present temperate state. He observed that the ground beneath our feet today becomes approximately 1°F warmer for every 50ft (15m) you dig down; made some assumptions about how much heat energy is held in warm rocks and how speedily it can pass through them; and decided the Earth is 20–400 million years old, a range he later narrowed down to 20–40 million years. Unfortunately, this result satisfied no one—it was too long for Biblical literalists and too short for the evolution of the profusion of life we now see on the planet. Yet it suited Kelvin, as he had already decided that the Sun had sufficient fuel to burn for only a further 20 million years. In fact, it was not realized until 1903 that the Earth's heat is continually replenished by internal radioactivity, and we now know the Sun is powered by a more potent mechanism than Kelvin realized—and nowadays we think both planet and star are roughly 4,500 million years old.

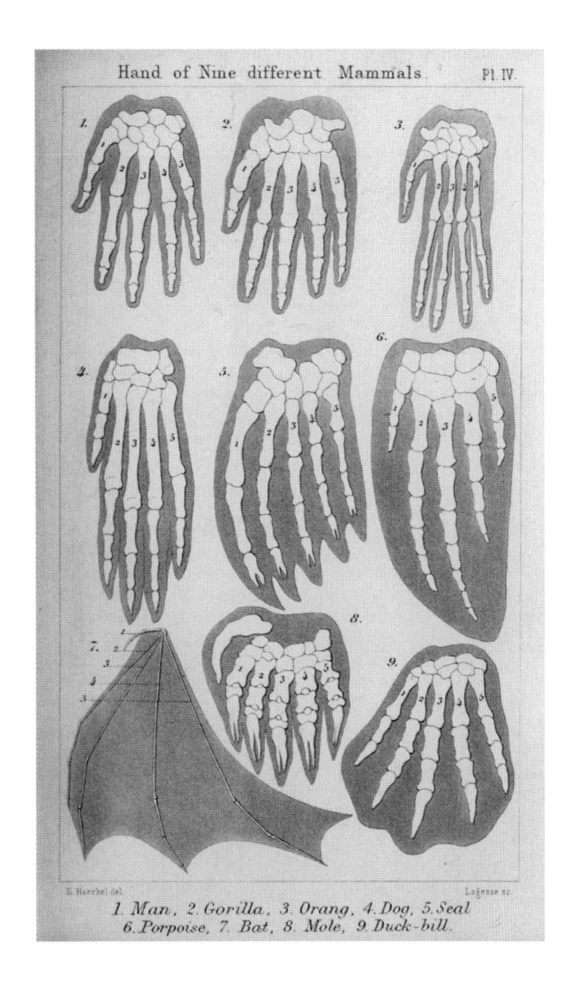


**Ernst Haeckel (1834–1919), *Natürliche Schöpfungsgeschichte*, vol. II, 1868; Hands of nine different animals.**

The context of paleontology was also changing in other ways. A major figure in late-nineteenth-century biology was Ernst Haeckel (1834–1919), professor of zoology at the University of Jena in Thuringia, Germany. Haeckel was a comparative anatomist and developmental biologist who became obsessed with natural selection, loudly promoting his own brand of it, "Darwinismus." Haeckel saw evolution not only as the way animals change over time, but also as something imprinted on each individual animal's life story. Most notably, his 1866 Theory of Recapitulation claimed that every animal recounts its evolutionary story as it develops in the womb or egg—thus, human embryos were claimed to pass through single-celled, wormy, fishy, reptilian, mammalian, and simian stages before they are born. While this idea now seems strange, it was immensely influential—and it even implied that paleontology is an "unnecessary" science because we can learn all we need to know about a species' evolution simply by observing its embryonic development. Not convincingly refuted until the 1920s, Recapitulation still looms surprisingly large in animal biology.

Across the Atlantic, paleontology was moving into an even more rambunctious conflict—the so-called "Bone Wars" of Charles Marsh (1831–1899) and Edward Drinker Cope (1840–1897). Recounted on pages 96–99, the acrimonious and destructive animosity between these two paleontologists started in the 1860s and continued for more than two decades until it left both men discredited and penniless. However, it also led to one of the most productive periods in fossil-hunting history, as new lodes were opened up in the American West, yielding more than one hundred new species of dinosaur alone. The specimens unearthed by these feuding paleontologists were to fill freshly built natural history museums across the United States and dominate public perceptions of ancient life, with famous *dramatis personae* such as *Allosaurus*, *Stegosaurus*, and *Triceratops*, and also the most famous of them all, the dubious invention that was *Brontosaurus* (see page 99). Titanic ancient mammals were also hewn from the rock, as well as fossil horses whose evolutionary progression from small to large has become a hackneyed trope of evolutionary science.

**Edward Drinker Cope (1840–1897), *Synopsis of the Extinct Batrachia, Reptilia, and Aves of North America*, 1869; Vertebra of *Elasmosaurus platyurus*.**

In 1891 paleontology first turned its attention to our own species. The Dutch anthropologist Eugène Dubois (1858–1940), a student of Haeckel, was so convinced there must be fossil evidence of "missing links" in human evolution that he traveled to the Dutch East Indies (now Indonesia) and discovered just such a link. From 1891, he mounted expeditions along the Solo River in Java, and near the village of Trinil discovered the top of a skull, a tooth, and two femurs of *Pithecanthropus erectus*, later renamed *Homo erectus*, but more often known as "Java Man" (see page 103). Java Man was a revelation to both scientists and public because it obviously fits somewhere along the evolutionary continuum between the great apes and humans—although not as neatly halfway as Dubois might have had us believe, what with his overegging of its ape affinities. Java Man fossils are probably 0.7–1.0 million years old, and have turned out to represent just one *Homo erectus* population of the many dispersed across the Old World between 1.6 and 0.7 million years ago. The evolutionary tree of hominins is now complex and tangled, but Java Man was the first twig discovered—and the twig that most radically changed how we see ourselves.

**Charles Walcott (1850–1927) with his son Sidney and daughter Helen at Burgess Shale Fossil Quarry, 1913.**

From the quarries at Solnhofen and the outcrops of the American frontier onward, rich fossiliferous seams were sometimes to become more famous than the paleontologists who excavated them, and even some of the creatures they disgorged. These *Lagerstätten* (German for "lair-places") may have gone down in paleontological folklore, but they betray a striking level of Eurocentrism and Americocentrism. A counterexample is the rarely mentioned Tendaguru Lagerstätte in what is now southern Tanzania. In 1906, a German mining engineer working in what was then German East Africa discovered some giant bones near Tendaguru Hill, and this led to large paleontological expeditions being mounted between 1909 and 1913 (see page 113). Over 200 tons of fossil bones were extracted, mainly by local workers, representing hundreds of species of plants, invertebrates, fish, amphibians, early mammals, and reptiles—dinosaurian and otherwise. Now on display mainly in Germany, the sheer profusion of these fossils has raised many questions about how they all ended up in one fossil bed—sudden mass deaths have been suggested, although the melancholy fossilization of entire dinosaur feet in an upright position has led some to suggest that the ancient Tendaguru swamp was a death-trap where generations of huge beasts became mired in mud only to starve or be slaughtered. Although Tendaguru boasts a dinosaur fauna just as spectacular as those found in the U.S., the names of its dinosaurs—*Kentrosaurus*, *Elaphrosaurus*, *Giraffatitan*—remain tellingly unfamiliar.

The discovery of arguably the most important fossil Lagerstätte of all came on August 30, 1909, when Charles Walcott (1850–1927; see page 118), Secretary of the Smithsonian Institution, stumbled upon a two-meter-thick shale bed in British Columbia. An expert on early invertebrates, Walcott had made annual fossil-collecting expeditions to the area around Mount Stephen for some years, usually assisted by his wife and children, but within two days of examining these new finds he knew they were especially important. We now know the rocks he found on the Burgess Pass—now called the "Burgess Shale"—are 509 million years old and thus contain fossils more than twice as old as any dinosaur. And the most remarkable thing about them is that they contain the detailed imprints of soft-bodied animals, from a time before bones and teeth even existed. The number and quality of the fossils was overwhelming—more than enough to keep Walcott busy for the rest of his life. However, it was not until the 1960s when they were reappraised by Harry Whittington, Derek Briggs, and Simon Conway-Morris of Cambridge University that it was realized just how important they are. The shales date from an era when animals were just starting to evolve into the many (twenty-five or so) body

**Reconstruction of *Anomalocaris canadensis* by Matteo De Stefano, 2016.**

plans we see today, a time of unparalleled evolutionary novelty now known as the "Cambrian Explosion." Although other fossils have since been found which date from this short, frenetic epoch of animal diversification, the Burgess Shale fossils remain unique in their scientific value and the sheer unlikeliness of their survival to the present day.

As this era of paleontological wonders drew to a close, it was geology which once more signaled the future. In 1912, Alfred Wegener (1880–1930) published his theory of *Kontinentalverschiebung*, or "continental drift," an idea eventually to prove as important to science as evolution or natural selection. Some scientists had already noticed the strange similarity between the outlines of certain continents—that between Africa's west coast and South America's east coast being the most striking example. However, it was Wegener who first proposed a narrative of continents slowly but inexorably gliding across the surface of the planet, and diverging to create oceans with huge, mid-oceanic mountain ridges. He argued it was actually the land masses' submarine continental shelves which tesselate together most neatly, rather than the coasts themselves, and he even suggested that all continents were once fused together in an ancient *Urkontinent* that we now call "Pangaea." Like Walcott, it would not be until the 1960s that the importance of Wegener's discovery was realized, in his case with the discovery of the mechanism underlying it: plate tectonics (see page 166). Continental drift is why we now realize that, not only are fossils the record of animals changing over immense stretches of time, but those evolving animals and their stony remains can undertake immense journeys across the face of the globe, riding their gigantic continent-rafts.

**Antonio Snider-Pellegrini (1802–1885), *La Création et Ses Mystères Dévoilés*, 1859; The Atlantic Ocean before the separation of the Americas and Africa.**

**Left: Hermann von Meyer (1801–1869), *Palaeontographica*, vol. X, 1861; *Archaeopteryx* fossil feather found by von Meyer at Solnhofen, Bavaria, Southern Germany, in 1861; and as drawn by von Meyer (right), from *Palaeontographica*, Cassel: Theodor Fischer, 1861, vol. X, plate VIII (detail).**

**Opposite: *Archaeopteryx lithographica* fossil (the "Berlin specimen"), found at Solnhofen, Bavaria, Southern Germany, in 1874.**

These images are of the first and third *Archaeopteryx* (ancient-wing) fossils discovered, both from the Solnhofen quarries in Germany. The limestone from this region is of an exceptionally high quality, laid down in fine sheets which may be peeled apart like the pages of a book—and occasionally one of those pages reveals something wonderful. The first fossil has been the more contentious of the two—doubts have been raised whether it belongs to *Archaeopteryx* or another dinosaur, and whether its asymmetrical shape proves it was a flight feather. The Berlin specimen, originally sold by its discoverer so he could buy a cow, is among the most complete and well preserved, although there have even been unfounded claims that the feather imprints were artificially etched onto the fossil.

**John Phillips (1800–1874),** ***Life on the Earth: Its Origin and Succession,*** **1860; Successive systems of marine invertebral life.**

The English geologist John Phillips extended the work of others to create an overarching scheme of geological chronology, with many of the subdivisions we still see on modern stratigraphical charts. On this diagram periods within the Paleozoic, Mesozoic, and Cenozoic Eras are correlated with the different invertebrate fossils borne by their strata.

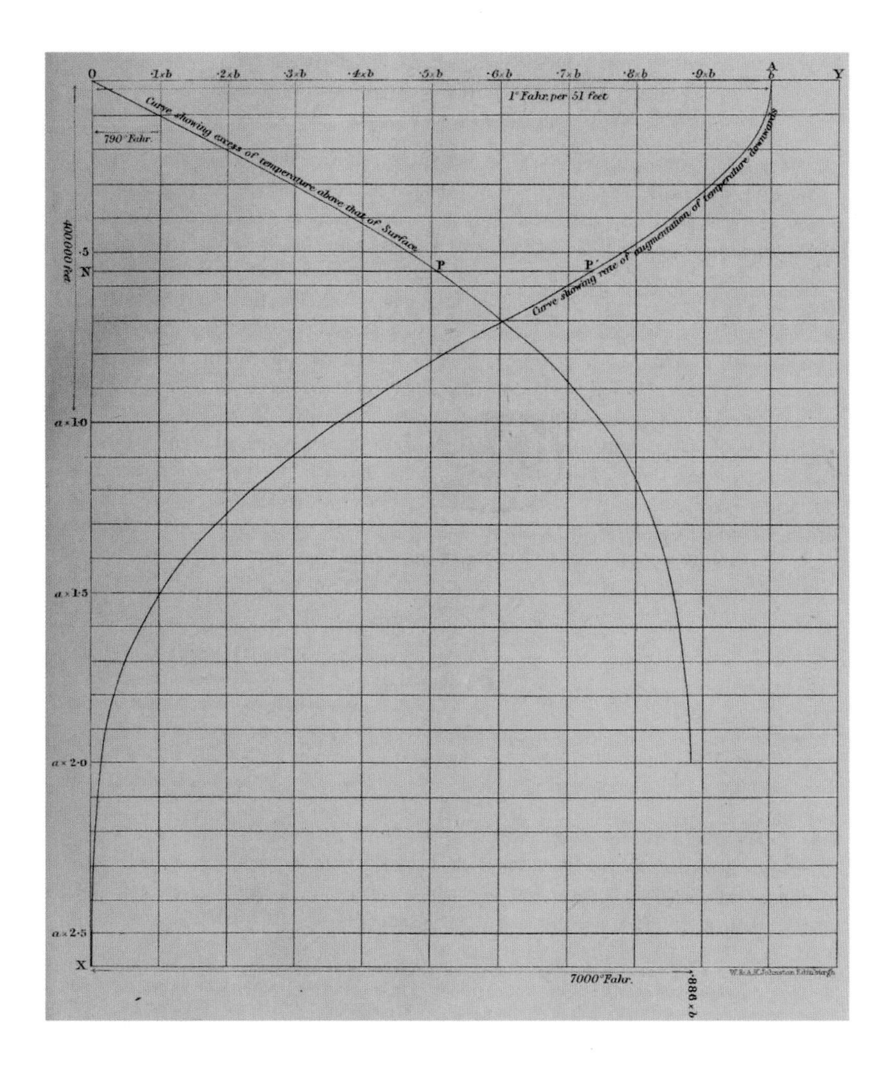


**William Thomson, Lord Kelvin (1824–1907), "On the Secular Cooling of the Earth," *Transactions of the Royal Society of Edinburgh*, vol. 23, 1861–1864; Increase of temperature downward in the Earth.**

Nineteenth-century advances in the science of thermodynamics, as well as some new mathematics, allowed the first modern quantitative estimate of the age of the Earth. Lord Kelvin made various assumptions—the initial temperature of the globe and the rate at which heat is conducted outward from its center and lost at its surface, along with measurements of temperatures at different depths under the ground—to calculate how long the Earth has existed. His estimate was 20–40 million years—far too long for those who believe in the absolute historicity of the Bible, but probably too short for evolution to have produced the diversity of life on Earth. Although Kelvin's result is indeed too low, it represents an honest scientific estimate, which erred only because it was not yet known that the Earth's heat has been partially replenished by its internal radioactive decay.

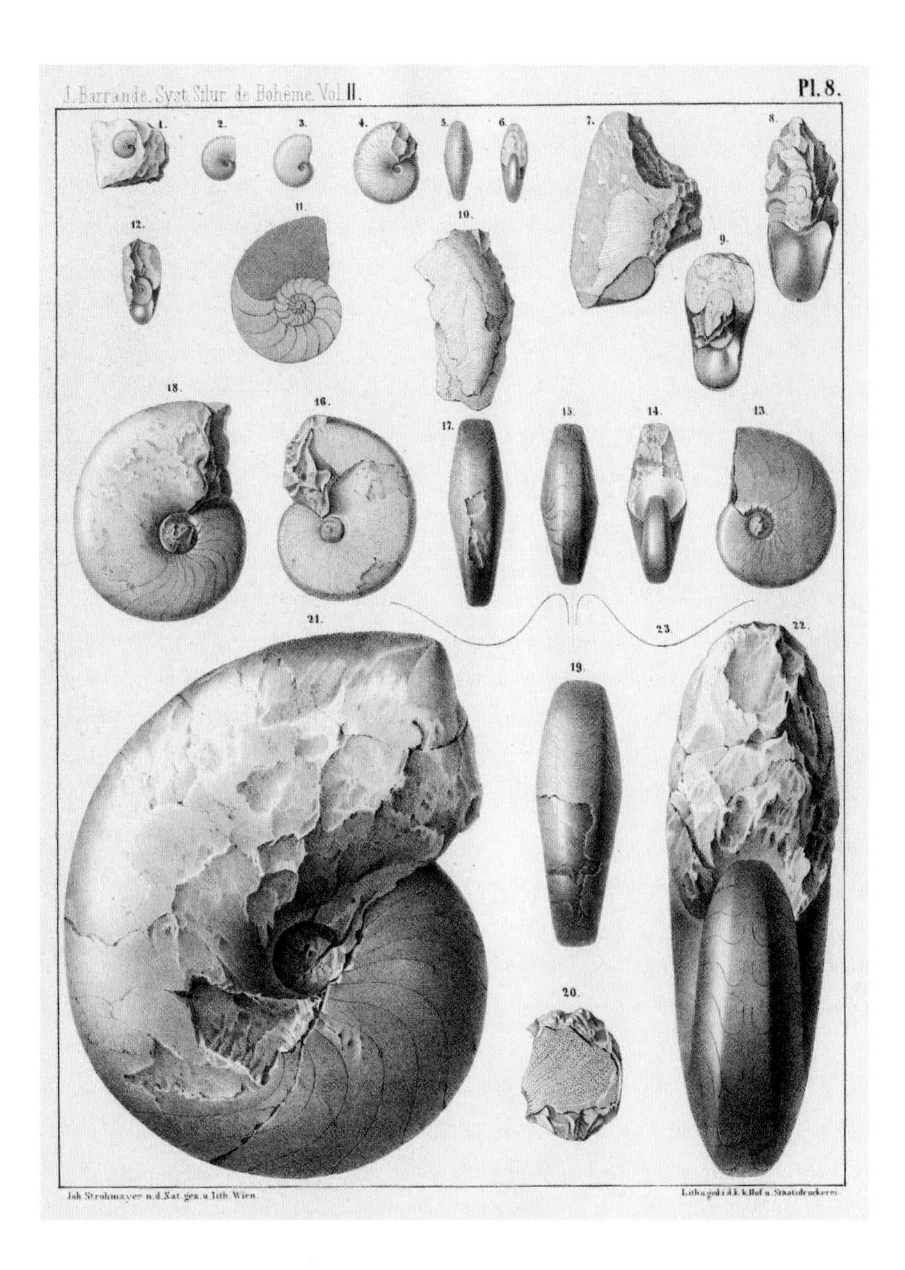


**Joachim Barrande (1799–1883),** ***Systême Silurien du Centre de la Bohême*****, 1852–1911, vol. 2, plate 8 (dated 1865); The ammonite** ***Goniatites fidelis*****.**

Working in Prague, Joachim Barrande's titanic multi-volume treatise on the fauna inhabiting what is now Bohemia during the Silurian Period (now defined as 444–419 million years ago) was crucial in the "expansion" of our knowledge of life to a date far more remote than the oldest dinosaur. Barrande described literally thousands of novel species, and the *Systême Silurien* proved to be too much for a single lifetime—a quarter of its bulk was published posthumously over a twenty-five-year period.

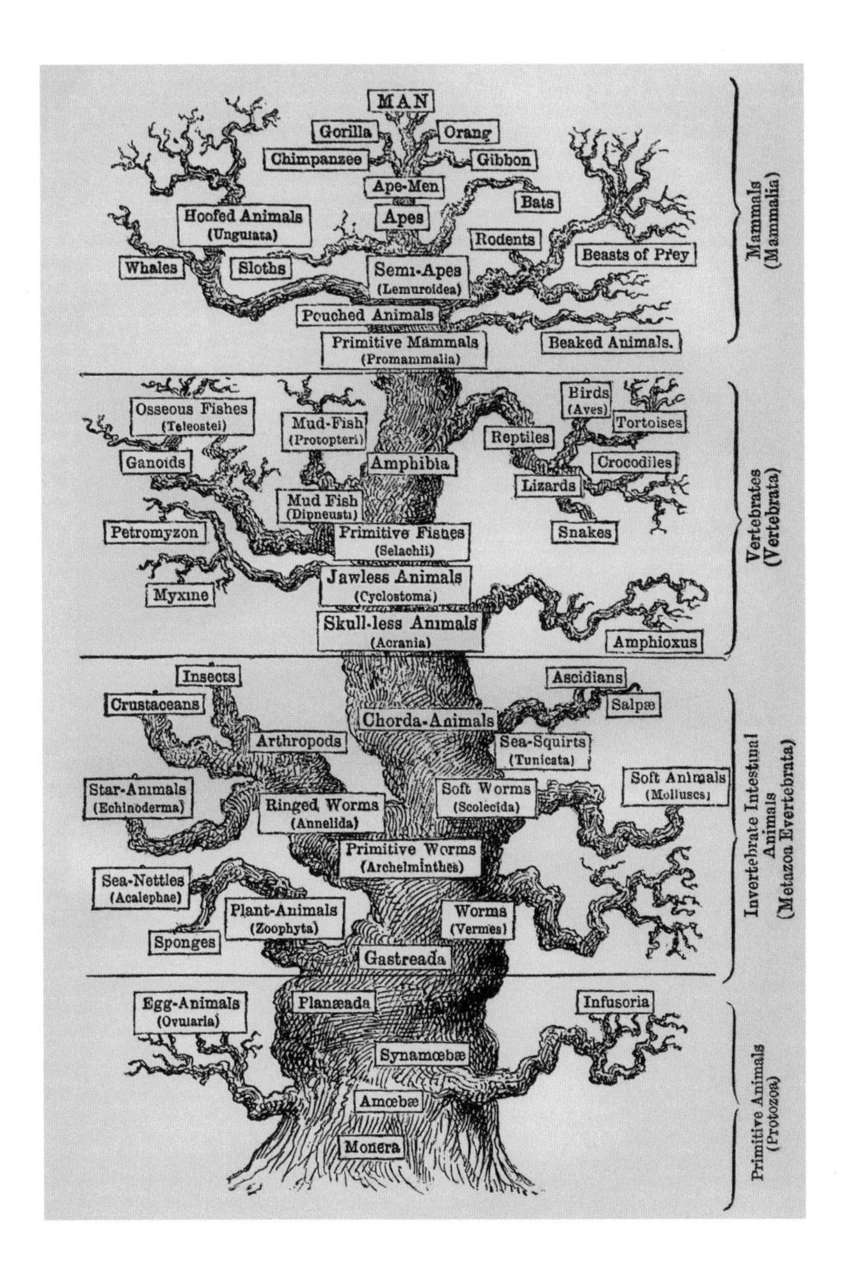


**Ernst Haeckel (1834–1919), *The Evolution of Man*, 1879; Pedigree of Man.**

Ernst Haeckel, at the University of Jena, was a proponent of natural selection and his excitable prose often describes Darwin in god-like terms. An intrepid traveler, pioneering embryologist, and prodigious writer, he was one of the best known and most widely read scientists of the time. Haeckel believed that evolution was a progressive process, guiding animals toward ever more complex and perfect forms, and drew direct parallels between evolution and the elaboration of developing embryos. Indeed, his obsession with embryonic development was so great that he largely neglected extinct species. This evolutionary tree, originally published in the 1874 German edition, *Anthropogenie oder Entwickelungsgeschichte des Menschen*, is just one of Haeckel's striking evolutionary trees—with humans proudly placed as the highest branch.

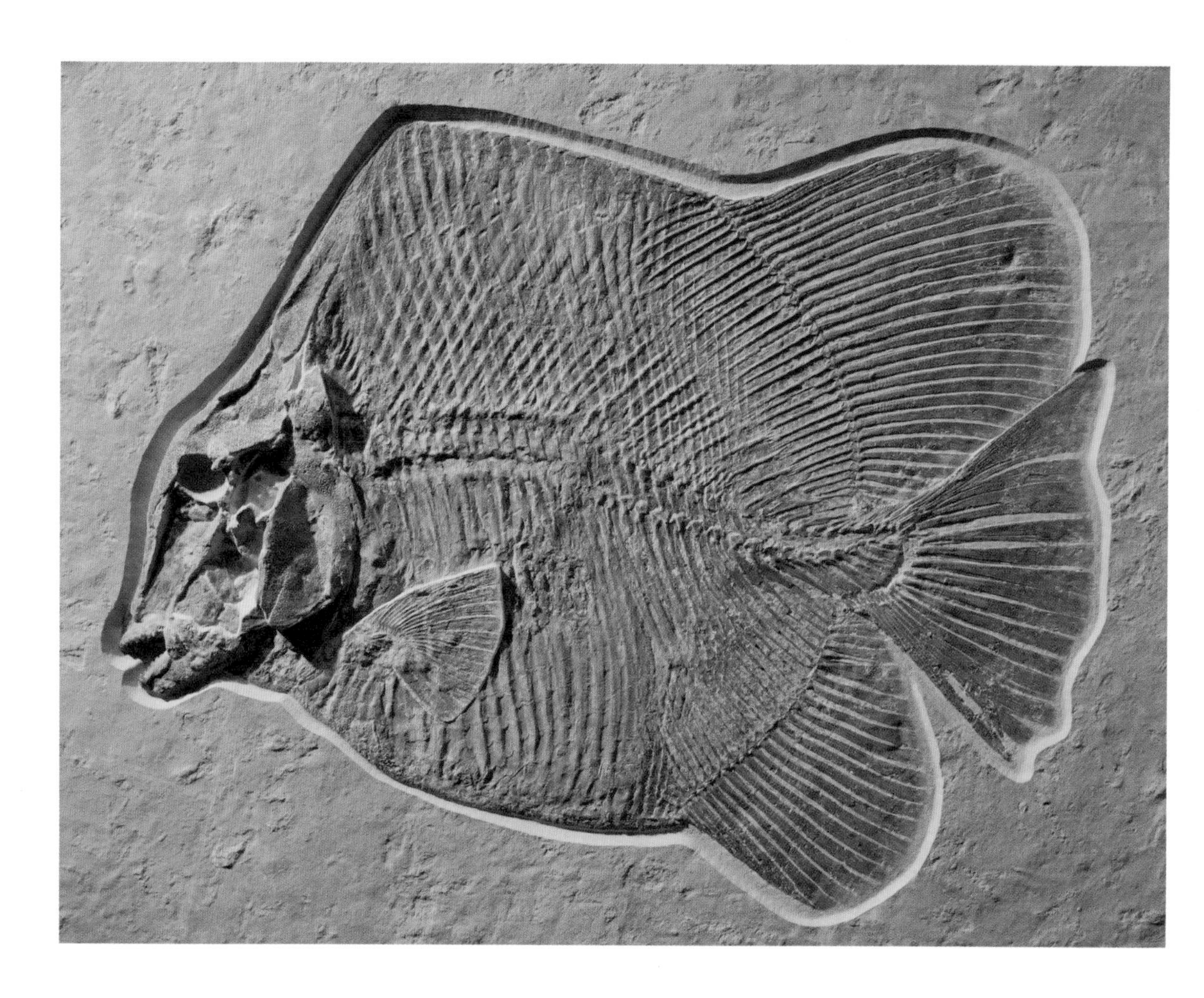

**Pycnodont fish *Macromesodon gibbosus* (Jurassic Period), found at Solnhofen limestone quarry, Bavaria, Southern Germany.**

The Solnhofen quarries hold far more fossils than just the famous *Archaeopteryx* (see page 86). Indeed, some of the most mundane creatures have left spectacular imprints in its limestone, including this medium-sized fish, a member of a widespread group.

THE DINOTHERIUM.

**Samuel Griswold Goodrich (1793–1860), *Johnson's Natural History*, vol. 1, 1868; *Deinotherium*.**

*Deinotherium* or "terrible beast" is one of the largest ever land animals, with a distinctive downturned lower jaw and downward-pointing tusks. As its trunk suggests, it was closely related to modern elephants, and may have been even larger. Inhabiting Europe, Asia, and Africa, *Deinotherium* species thundered across the plains between 25 and 1 million years ago.

**Lithograph by Paul Stewart, after Benjamin Waterhouse Hawkins (1807–1894), *Diagrams of the Extinct Animals*, 1862; Dinosaurs and pterosaurs.**

A Victorian summary of "British dinosaurs," this montage depicts *Iguanodon*, *Hylaeosaurus*, and *Megalosaurus*, whose appearance and posture are depicted with varying degrees of accuracy, as well as some pterosaurs looking on menacingly.

3
4
2
1
7

# CHARLES MARSH (1831–1899) AND EDWARD DRINKER COPE (1840–1897)

## The Bone Wars

Paleontology may not seem a likely arena for internecine struggle, but old bones can elicit strong emotions, and 1869 marks the start of the angry but productive "Bone Wars" between Charles Marsh of the Yale Peabody Museum and Edward Drinker Cope of the Philadelphia Academy of Natural Sciences.

Marsh briefly worked alongside Cope at his excavation sites in Haddonfield, New Jersey, but their relationship understandably soured when Marsh bribed the local quarry owners to send all future fossil finds to him. Thus commenced more than twenty years of the most competitive, successful, and acrimonious fossil hunting in history, played out mainly in the American West.

**Edward Drinker Cope (1840–1897), posing here with the skull of the extinct large, horned mammal *Eobasileus*, 1872.**

Cope initially went west with Ferdinand Hayden's U.S. Geological Survey of Colorado and Wyoming, famously exploring the Yellowstone region, while Marsh's expeditions were backed by Yale University, whose Museum of Natural History had been largely funded by his uncle. Things soon became competitive, with both men bankrolling large teams of fossil hunters, and tactics veered toward theft, bribery, slander, espionage, and vandalism. It is even claimed that both men would order the destruction of remaining fossils as they left excavation sites—a scorched earth policy intended to frustrate their competitor. Both men would publish minor variants in skeletons as entirely new species to increase their tally, but occasionally they would unite in undermining other paleontologists who announced discoveries before them.

Yet their struggle was undoubtedly productive, as now-famous Lagerstätten, including Dinosaur Ridge, Morrison, Colorado, and Como Bluff, in Wyoming, disgorged their riches. Between them Cope and Marsh

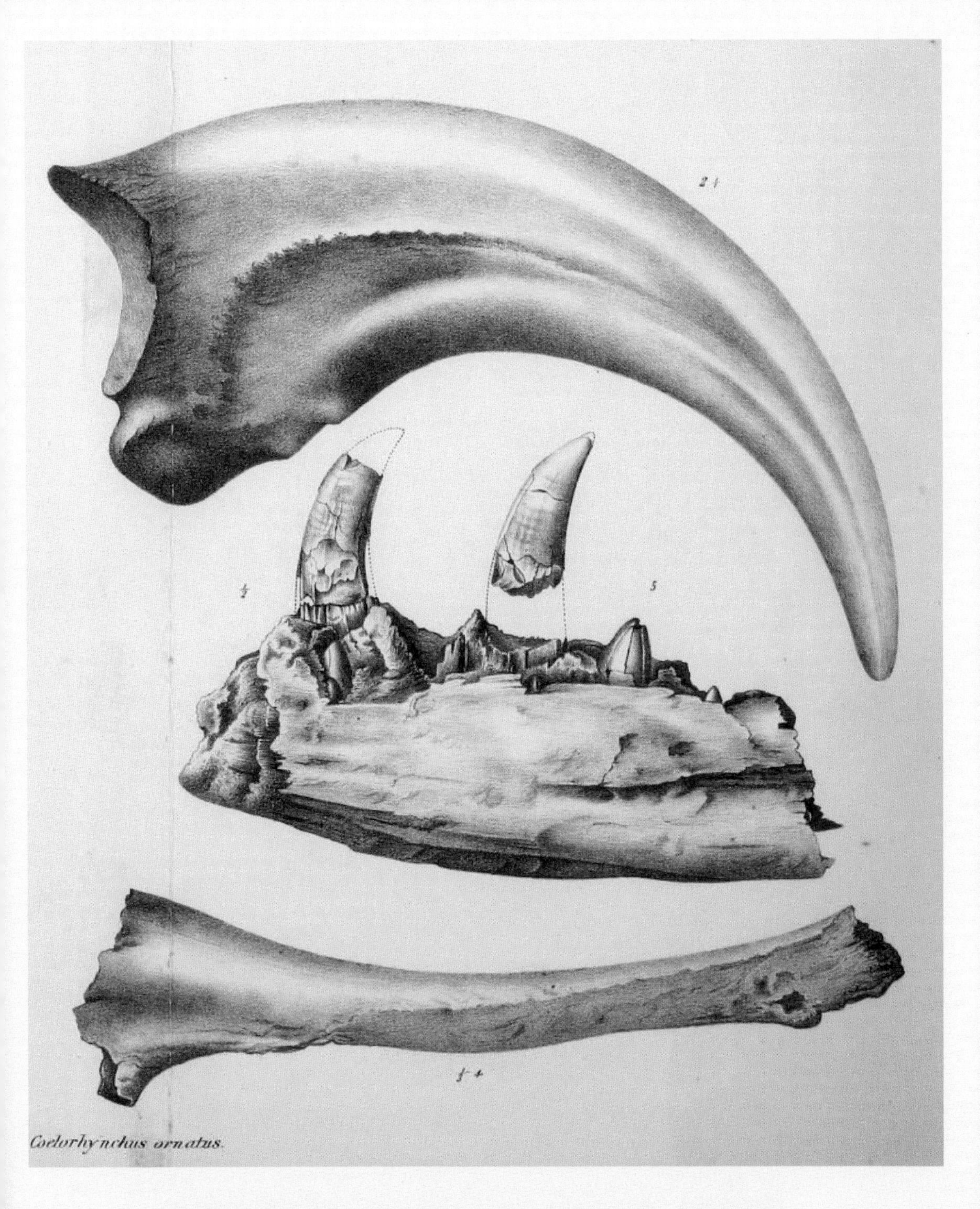


**Edward Drinker Cope (1840–1897),** ***Synopsis of the Extinct Batrachia, Reptilia, and Aves of North America*****, vol. 1, 1869;** ***Laelaps aquilunguis*** **(now** ***Dryptosaurus*****).**

**The Yale University expedition party, with Charles Marsh standing in the center of the back row, in 1872.**

described many of the most famous Jurassic and Cretaceous dinosaurs—*Stegosaurus*, *Allosaurus*, and *Triceratops*—as well as giant fossil mammals and the plesiosaur *Elasmosaurus*, which Cope in his haste reconstructed with its skull on the tip of its tail.

One of the most famous casualties of the fog of paleontological war was poor old *Brontosaurus*. A species called *Apatosaurus* had already been described from incomplete skeletons, and indeed had been displayed in museums erroneously bearing the head of another giant sauropod (see opposite). When a more complete skeleton was discovered it was dignified with the exciting new name "thunder-lizard," or Brontosaurus. However, it was nothing more than another *Apatosaurus* skeleton, and even though the mistake was discovered in 1909, the non-existent, newly named species "Brontosaurus" gripped the public imagination and survived in museums until the 1970s.

Thus, these two men, who so hated each other, are now usually spoken of in the same breath: "Cope and Marsh." Scientific ego and the financial rewards of exploiting a new frontier's underground wealth were too much for both, leading to ruin within their own lifetimes and the most ignominious of scientific legends.

*Apatosaurus* (right) with the skull from a *Camarasaurus*-like species, 1934, Carnegie Museum of Natural History.

**Watercolor by Arthur Lakes (1844–1917) showing Yale University expedition members E. Kennedy and Bill Reed with dinosaur bones at Como Bluff, Wyoming, circa 1879.**

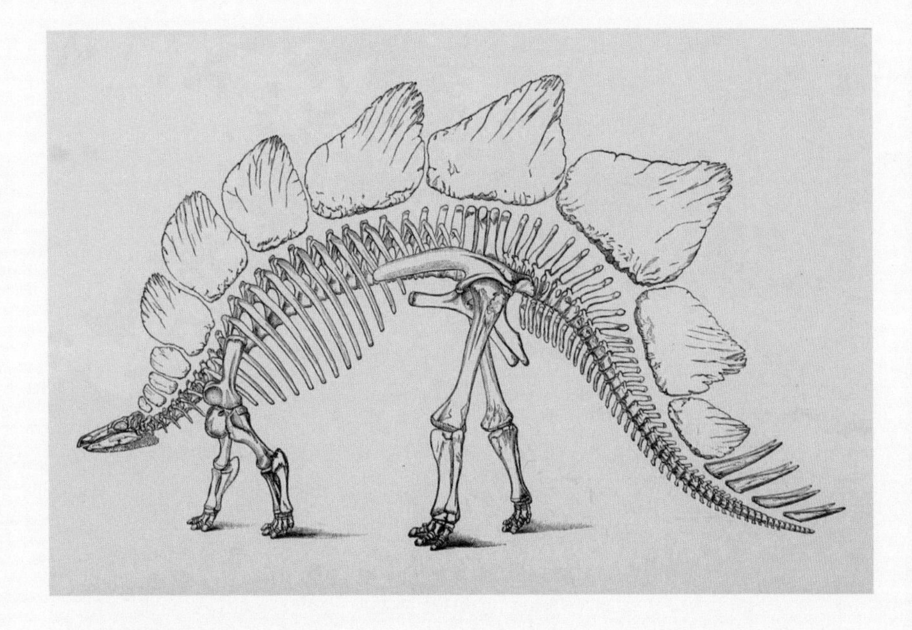

**Charles Marsh (1831–1899), *The Dinosaurs of North America*, 1896; *Stegosaurus ungulatus*.**

**Dinosaur tracksite from the Lower Cretaceous Period at Dinosaur Ridge, Morrison, Colorado.**

One of the sites exploited during the nineteenth-century "Bone Wars," Dinosaur Ridge's exposed sandstones bear fossilized trackways of the bipedal dinosaur *Caririchnium*. Trackways may seem rather transient things to survive to the present day, but fossilization is a matter of probabilities, and an animal leaves far more footprints during its lifetime than it leaves bones after it dies. Trackways have also been used to derive information about animals' locomotion and behavior.

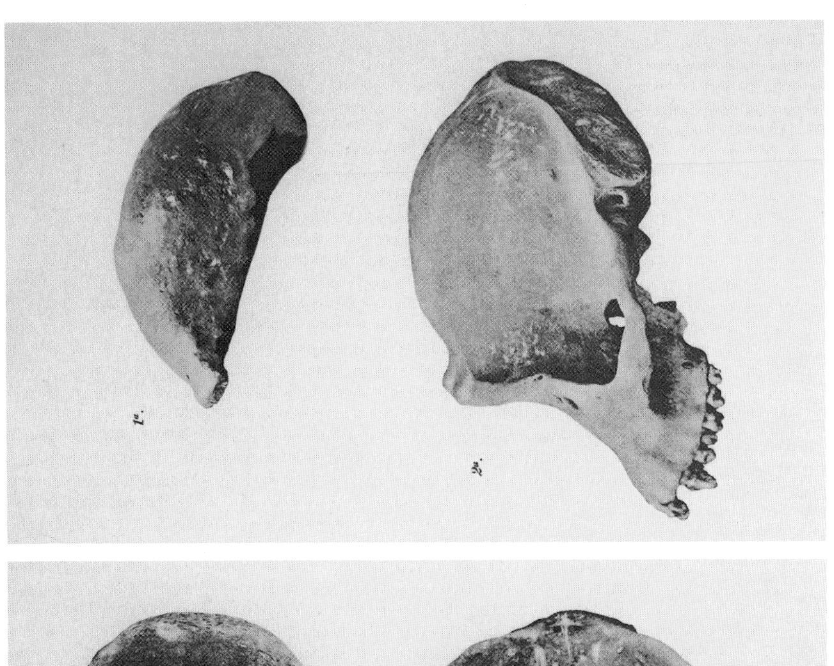


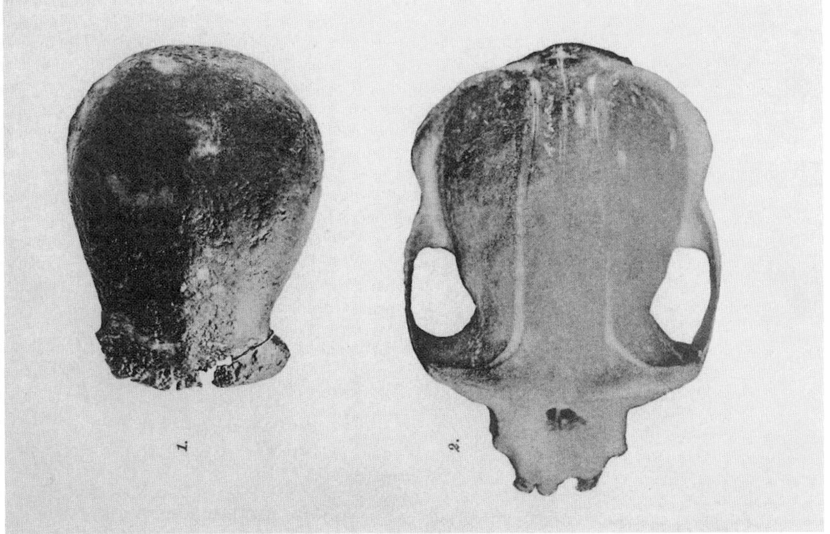


**Eugène Dubois (1858–1940), *Pithecanthropus erectus*, 1894; Skull of Java Man.**

Dating from 1.0–0.7 million years ago, "Java Man" was discovered in the early 1890s by a team led by Eugène Dubois, one of a new breed of paleontologists keen to demonstrate evidence of human evolution in the fossil record. Probably derived from three different individuals, the finds included the top of a skull, teeth, and a very "modern-looking" thighbone. At the time, Java Man was the oldest known hominin, and given the name *Pithecanthropus*, but it has since been welcomed into the fold of *Homo erectus*—emphasizing a closeness to modern humans which Dubois himself underplayed, preferring to think of these fossils as being "halfway" between humans and apes. By pouring mustard seeds into the cranium, its volume was calculated to be approximately 800 milliliters, satisfyingly intermediate between chimpanzees and humans (400 and 1,400 milliliters, respectively).

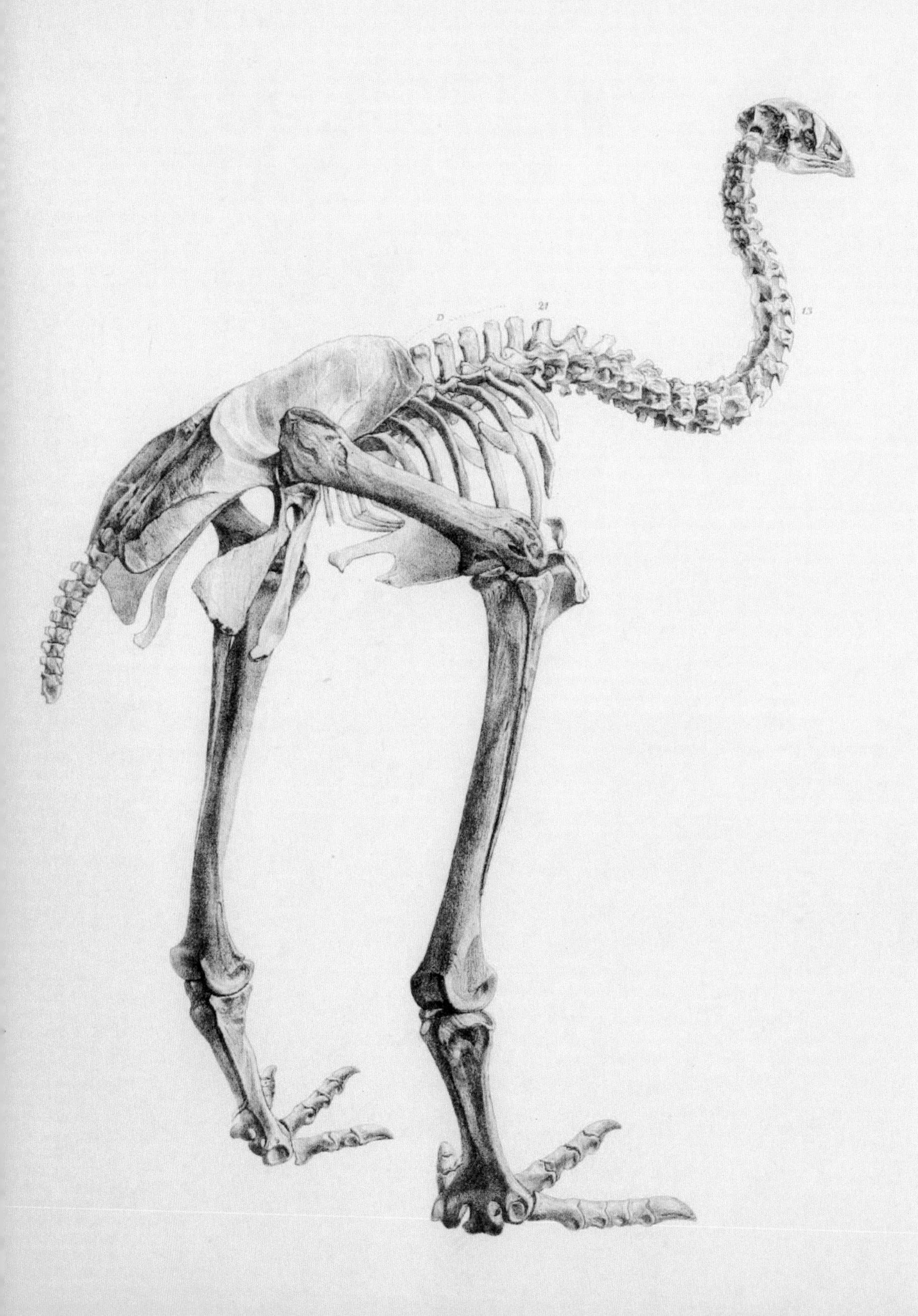
D
21
13

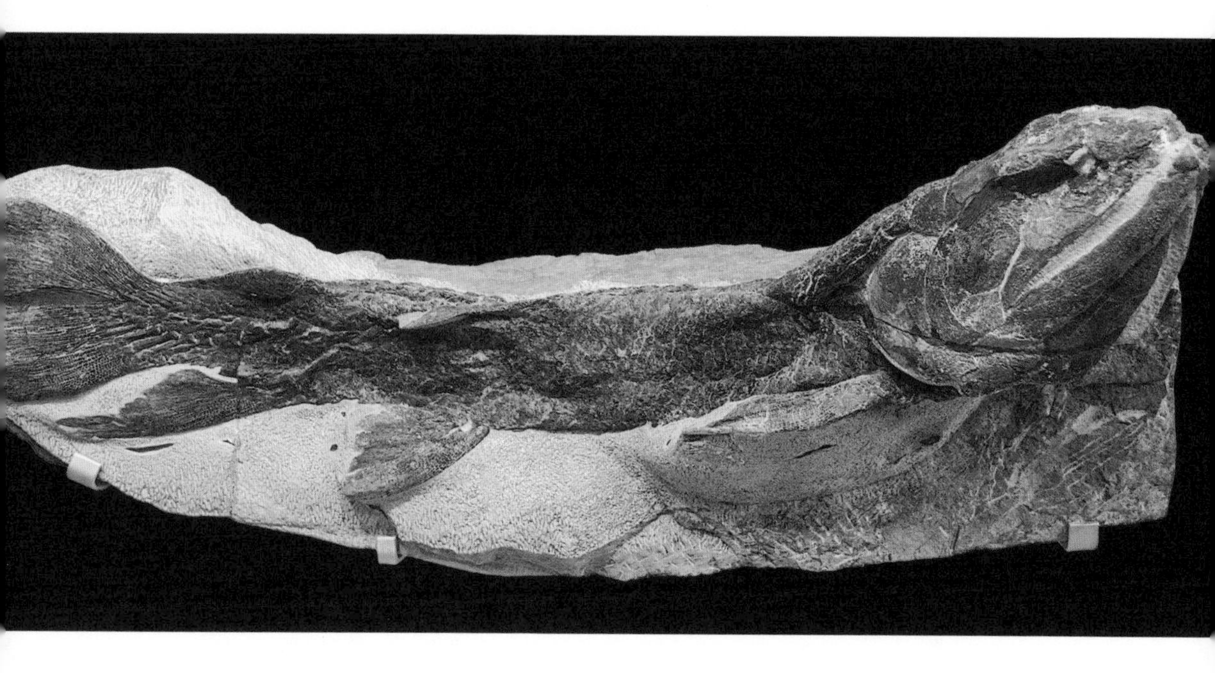

**Devonian Era fish *Eusthenopteron foordi*, found at the Escuminac Formation, Miguasha, Quebec, Canada.**

Discovered in 1881, *Eusthenopteron* (strong-fin) is celebrated mainly because of its relatives—while obviously fish-like in appearance and habit, this lobe-finned fish is often cited as the "classic" close relative of tetrapods (land vertebrates). The *Eusthenopteron* genus contains several species dating from approximately 380 million years ago, of which thousands of specimens have been discovered, mainly in Canada. Its teeth and nostrils are similar to tetrapods, suggesting that some characteristics we usually associate with land-living vertebrates actually evolved some time before the transition to land.

**Opposite: Lithograph by James Erxleben, *Transactions of the Zoological Society of London*, vol. XI, 1880; Little bush moa (now *Dinornis parvus*).**

Among the largest birds ever to exist, there were several species of giant flightless moa in New Zealand when humans arrived. By 600 years ago all were gone. The origins and antiquity of the creatures are unclear; unlike ostriches and emus they had entirely lost their wings, and locomotion was by means of their robust legs.

**Fossil cycad *Sphenozamites*, Jurassic Period.**

In 1892 the first fossil cycad was discovered in what was to become, between 1922 and 1957, the short-lived Fossil Cycad National Monument in South Dakota. Often confused with palms, cycads are non-flowering plants which still exist today, but were once far more abundant and diverse, dating back 300 million years or more. It is likely they were to some extent out-competed by faster-growing flowering plants (angiosperms) which spread rapidly in the Cretaceous Period.

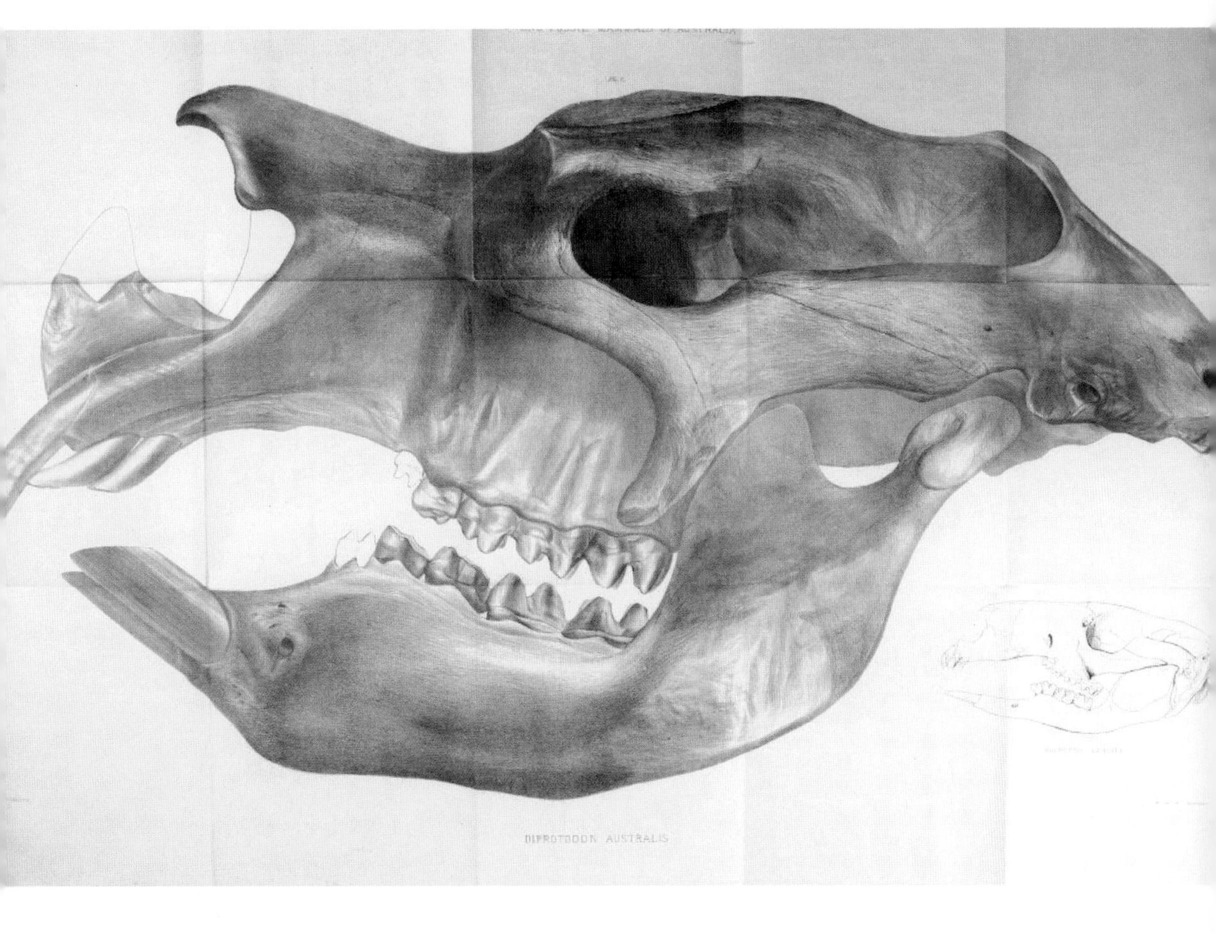


**Richard Owen (1804–1892),** ***Researches on the Fossil Remains of the Extinct Mammals of Australia*****, 1877;** ***Diprotodon australis*** **skull and jaw.**

Identified and characterized in the mid-to-late nineteenth century, the two-ton giant wombat *Diprotodon* is the largest known marsupial ever to have existed. Like its smaller modern relatives, *Diprotodon* had gnawing teeth and forepaws adapted for digging. It must have been a destructive creature, although it was almost certainly too large to create burrows—modern wombats are, in fact, the largest burrowing animals in the world. In existence 1.5–0.044 million years ago, it is speculated that its demise was a result of human hunting or environment modification. *Diprotodon*'s fossil bones may be the origin of the Aborigine legend of the "bunyip," a large, ferocious, swamp-dwelling beast.

# CHARLES KNIGHT (1874–1953)

## *Leaping Laelaps* and its Co-stars

Charles Knight is perhaps the most influential paleontological artist of all time. A commercial artist, he created many paintings for the American Museum of Natural History, in New York, the Natural History Museum of Los Angeles County, and the Yale, Carnegie, and Smithsonian Museums, as well as creating a spectacular series of murals for the Field Museum of Natural History in Chicago.

Throughout Knight's career many scientists were dismissive of his credentials as a paleontologist, but the vivacity and impact of his work has outlived those criticisms, and indeed those critics. With a century's hindsight it is, of course, easy to find fault with the technical aspects of his ancient creatures—for example, some of his larger dinosaurs look gray and sluggish and are shown in amphibious poses, apparently necessary to buoy

**Charles Knight (1874–1953), *Dimetrodon* with *Edaphosaurus* behind, 1897.**

Charles Knight (1874–1953), *Leaping Laelaps*, 1897.

up their bulk. Similarly, bipedal dinosaurs often rest partly on their tails, as if a "tripod" stance is the only way they could stand.

His most famous painting, however, depicts the *Tyrannosaurus*-like *Laelaps* (see page 97)—now renamed *Dryptosaurus*—in a very different way. Named after the dog of Greek mythology who always caught his quarry, the 1897 *Laelaps* is shown as an agile, speedy creature of the sort not seen again in dinosaur depictions until the "Dinosaur Renaissance" of the 1960s.

Despite the scientific lapses in some of his pictures, Knight's influence on our conception of the ancient world has lasted to the present day. Certainly, he was pivotal in cementing our disconcerting realization that these alien and fearsome animals once stalked the very same ground on which we now stand.

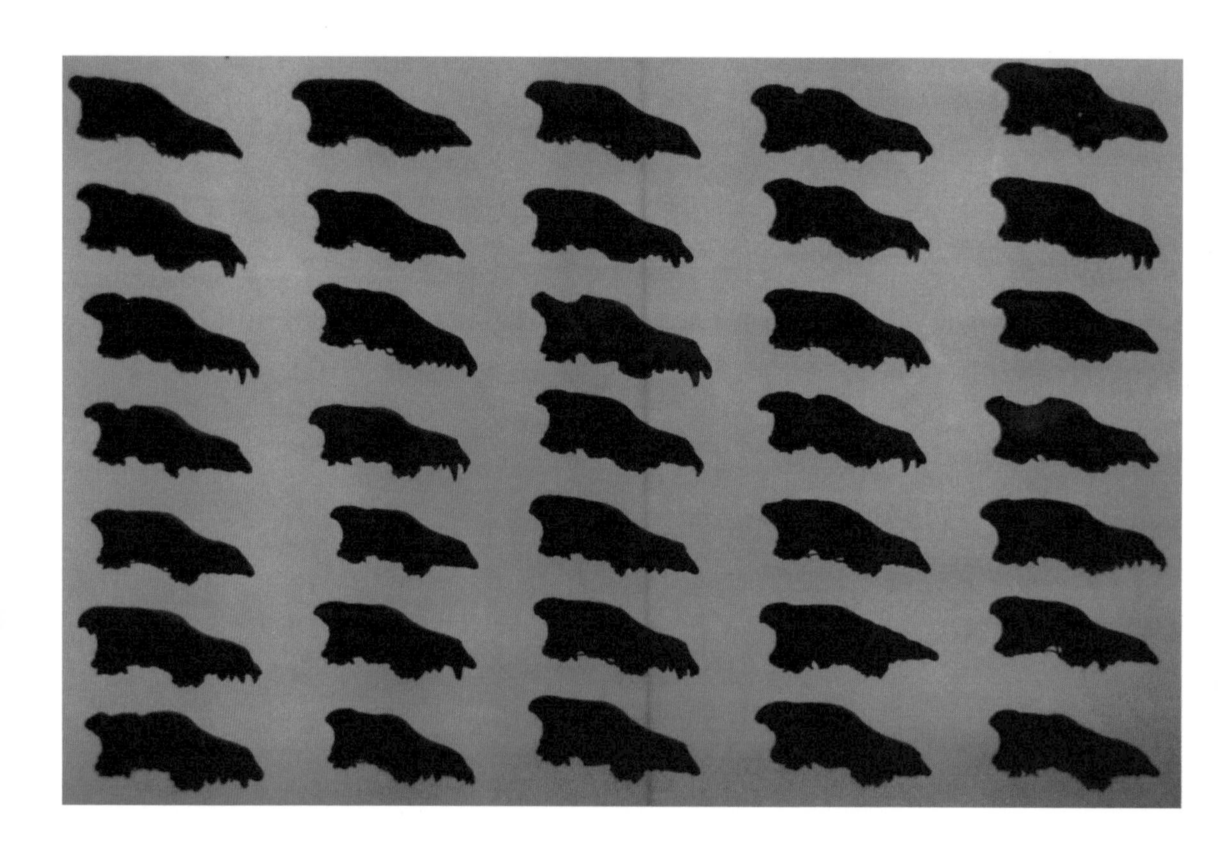

**Photograph by the author, showing dire wolf skulls found at La Brea Tar Pits, Los Angeles, 2015.**

Now hemmed in by the sprawl of Los Angeles, the La Brea Tar Pits acted like fly paper for large and small members of the Pacific coastal fauna between 50,000 and 10,000 years ago (0.05–0.01 million years ago). Over three million fossils have been found, and the most spectacular are those of the large mammals whose bones are found in jumbled collections in the pits. The "tar" is actually asphalt, the densest fraction of crude oil, and it is assumed that large herbivorous mammals became stuck in it and attracted the unwelcome attention of carnivores who then became mired themselves. As a result, the most numerous mammalian remains are those of the dire wolf, and this image is of a small fraction of a spectacular display of some of the four thousand wolf skulls found in the pits.

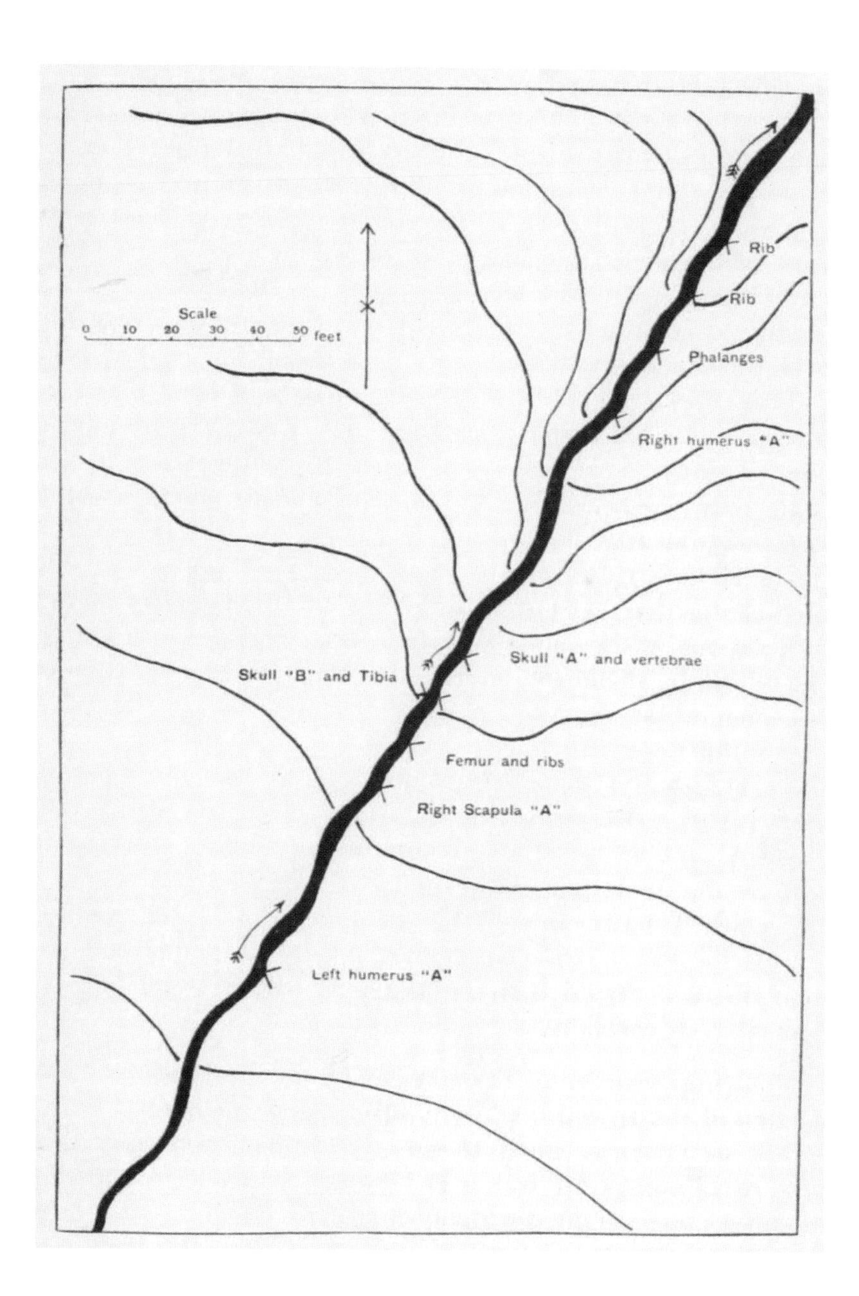


**Wynfrid Duckworth (1870–1956), *Studies from the Anthropological Laboratory, The Anatomy School, Cambridge*, 1904; The dispersive power of running water on skeletons.**

Very few organisms become fossilized after they die—only on the rare occasions when a certain combination of specific conditions is met. For land animals, the flow of surface water has important and sometimes surprising effects, leading to selective deposition and preservation of particular species, and even particular bones from an individual species.

***Triceratops elatus* skull (with bones from other dinosaurs), found by Charles Marsh in Niobrara County, Wyoming, 1891.**

In 1891 Charles Marsh (see page 96) discovered a particularly spectacular skull from a new species of the genus *Triceratops*, and proceeded to mount it upon a skeleton assembled from bones of a variety of related species—an assembly called, somewhat euphemistically, a "composite mount."

**Top: Excavating *Giraffatitan brancai*, Tendaguru expedition, 1909–1913.**
**Above: local chief supervisor Boheti with the giant rib of a *Giraffatitan brancai*.**

The Tendaguru fossil beds lie in what is now southern Tanzania, and between 1909 and 1913 expeditions led by the German colonial power yielded hundreds of tons of fossil invertebrates and plants, as well as bones from a wide variety of backboned animals, some enormous. The site has now been divided into over a hundred distinct fossiliferous areas, yet has yielded few complete skeletons, which may be one of the more benign reasons why it is less famous than some North American and European Lagerstätten.

**Barnum Brown (1873–1963) with a mounted *Pteranodon*, 1938, American Museum of Natural History.**

Barnum Brown (see page 129) was a paleontologist and fossil hunter whose work for the American Museum of Natural History in New York spanned many decades, and in these photographs he is shown with two of the most famous Mesozoic reptiles. *Pteranodon* (above) inhabited the Western Interior Seaway, a Cretaceous Period marine ecosystem stretching from north to south across what is now the center of North America. Its wingspan of 23ft (7m) was for a long time the largest known. This mount demonstrates how pterosaurs' wings were largely suspended from an enormously enlarged single finger—digit IV, or "ring finger." Opposite, *Tyrannosaurus rex* needs no introduction, and this specimen was one of the most complete known at the time the photograph was taken.

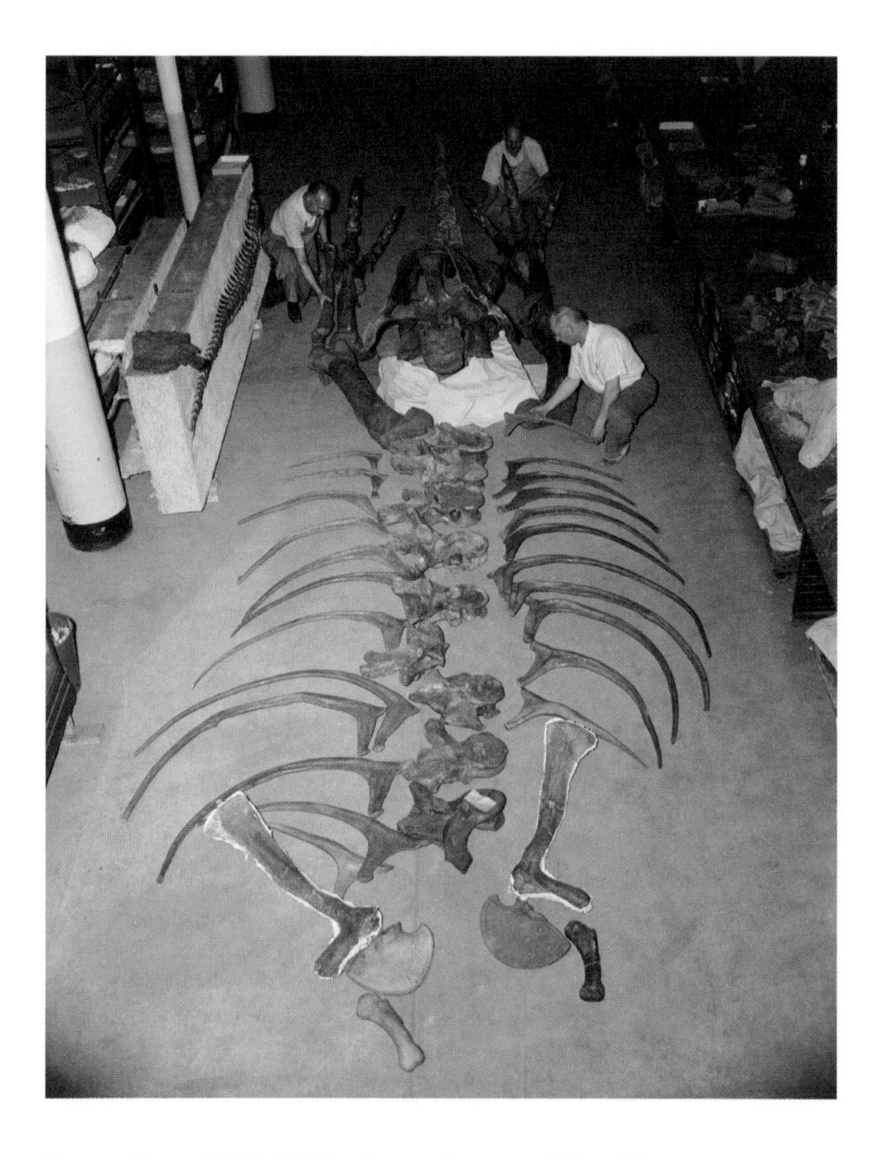

**Barnum Brown (1873–1963) and co-workers assembling a *Tyrannosaurus rex* skeleton, 1942, American Museum of Natural History.**

**Overleaf: John Bell Hatcher (1861–1904), *The Ceratopsia*, 1907; Skull of *Triceratops serratus*.**

Hatcher, a colleague of Charles Marsh (see page 96), discovered two genera of Cretaceous horned and frilled "ceratopsian" dinosaurs, the famous *Triceratops* and the even larger *Torosaurus*. Among the austere *Monographs of the United States Geological Survey* hides volume 49, Hatcher's *Ceratopsia*, full of stunning engravings of huge and ornate skulls, hewn from frontier rock.

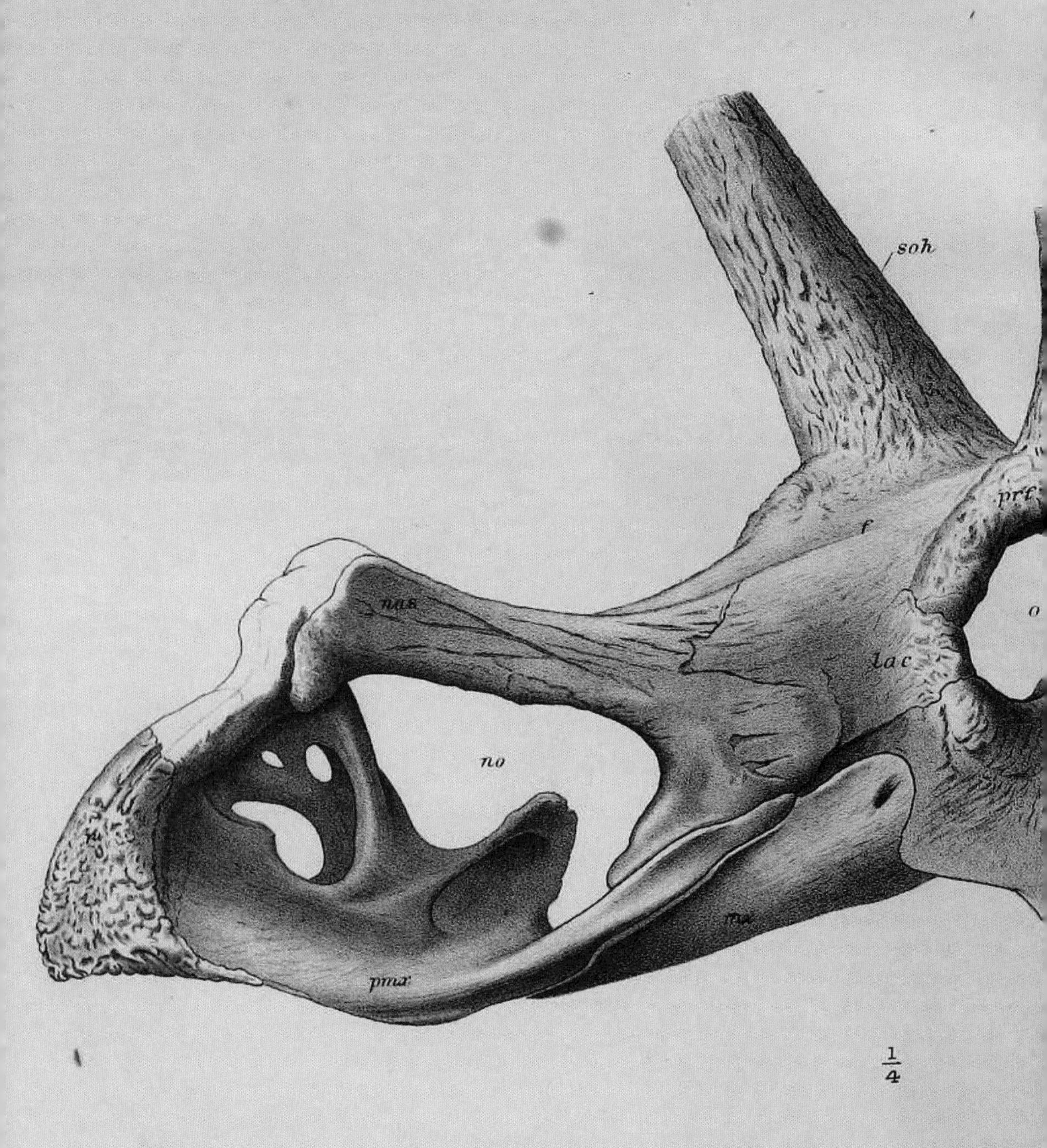
soh
prf
f
nas
o
lac
no
mx
pmx
1/4

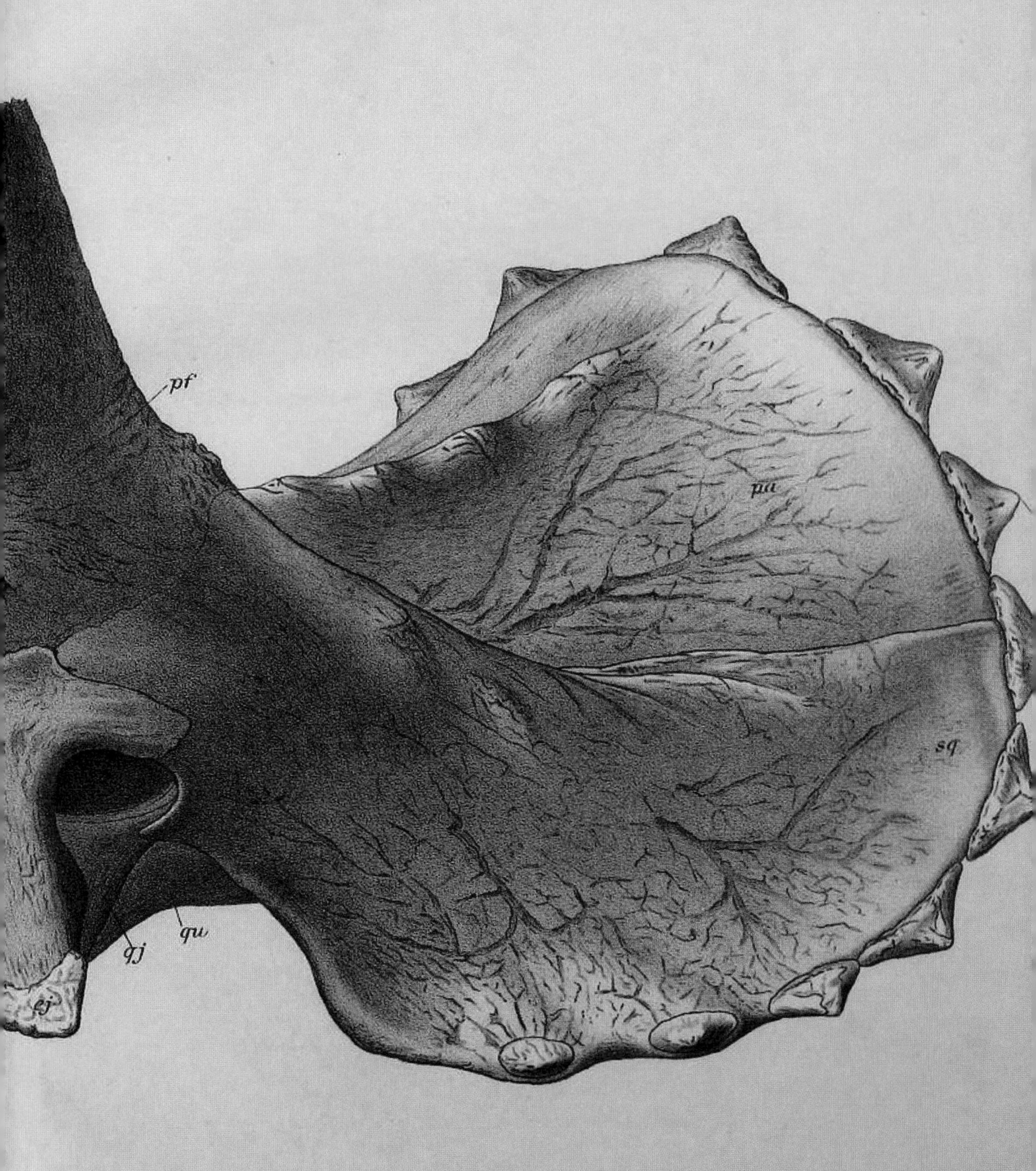


SERRATUS Marsh

# CHARLES WALCOTT (1850–1927)

## The Burgess Shale and the Cambrian Explosion

Charles Walcott's name will forever be linked to a few-meters-thick seam of shale high on a mountain ridge in British Columbia. Over the course of several fossil-hunting seasons between 1908 and 1924, the Burgess Shale, as it is now known, gave up hundreds of fossils of an amazing level of preservation, including vestiges of soft tissues. The fossils date from a time 508 million years ago when multicellular animals were diversifying into an array of novel forms, including the major groups we see today—a seminal epoch now called the "Cambrian Explosion."

**Right: Drawing by Elvira Wood after Charles Walcott's notes, *Memoirs of the Connecticut Academy of Arts and Sciences*, vol. 6–7, 1810.**

**Opposite: Complete fossil specimen of *Marrella splendens*, showing antennae and appendages, found at the Walcott Quarry, Burgess Shale, British Columbia, in 1912.**

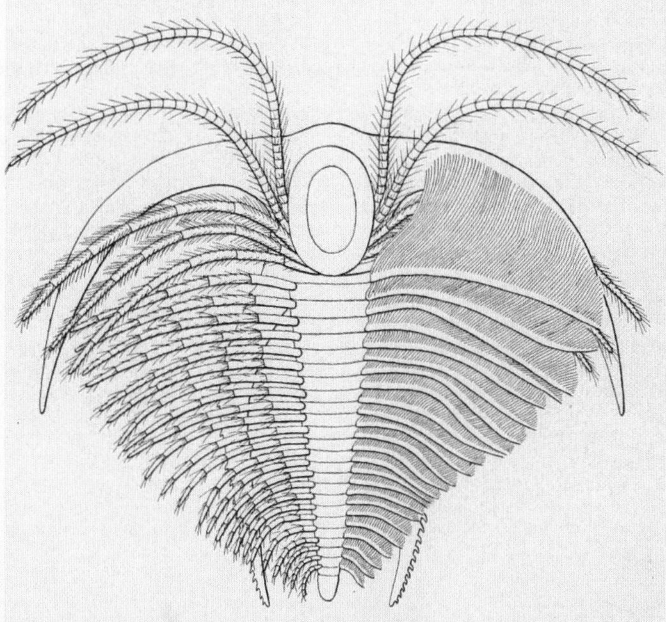

**Charles Walcott (1850–1927) pictured (with hammer) excavating with others at the Burgess Shale, British Columbia, in 1912.**

The region around Mount Stephens became known to Walcott as early as 1887 when he authored a paper on trilobites found in the region by others, but it was not until 1907 that he ventured there himself. In his notes he describes riding a pony up the slopes from the nearby railway halt, excavating throughout the day, and then leading his freshly laden beast back down the mountainside. Often he excavated with his wife and family—and he left an excellent photographic record of his findings as well as charming images of the communal process of excavation.

The 1907 expedition was concerned mainly with establishing the geology and stratigraphy of the region, and fossil excavation began in earnest in 1908. It was summer 1909 when Walcott first mentions in his notebook that he has found an exciting new seam containing unusually well-preserved fossils of sponges and crustaceans, many with evidence of soft tissue preservation. Summer excavations were to become a regular

family outing; so many fossils were found that Walcott did not have time to fully study and publish accounts of many of them.

Frustratingly, for decades after Walcott's death the Burgess Shale fossils lay largely unstudied in museum drawers across the United States and it was only in the 1960s that attention was once more focused upon them, initially by Alberto Simonetta and Laura Delle Cave of the University of Camerino in Italy. Thereafter, it soon became clear that the Shale fossils were even more important than Walcott had realized, representing a geological snapshot of a time when complex life on Earth had only just begun.

**Fossil specimen of *Pikaia gracilens* found by Charles Walcott at the Burgess Shale, British Columbia, in 1911.**

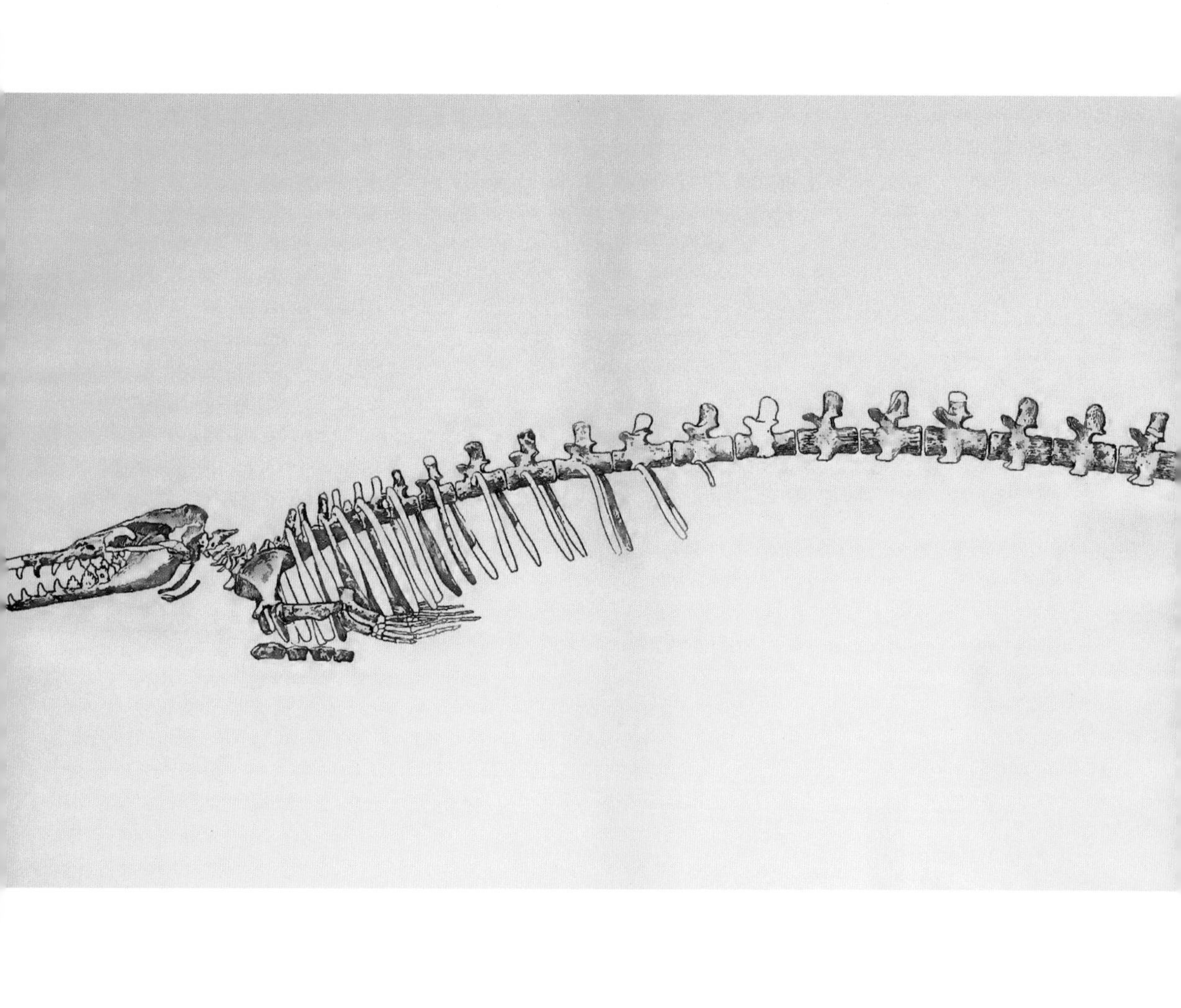

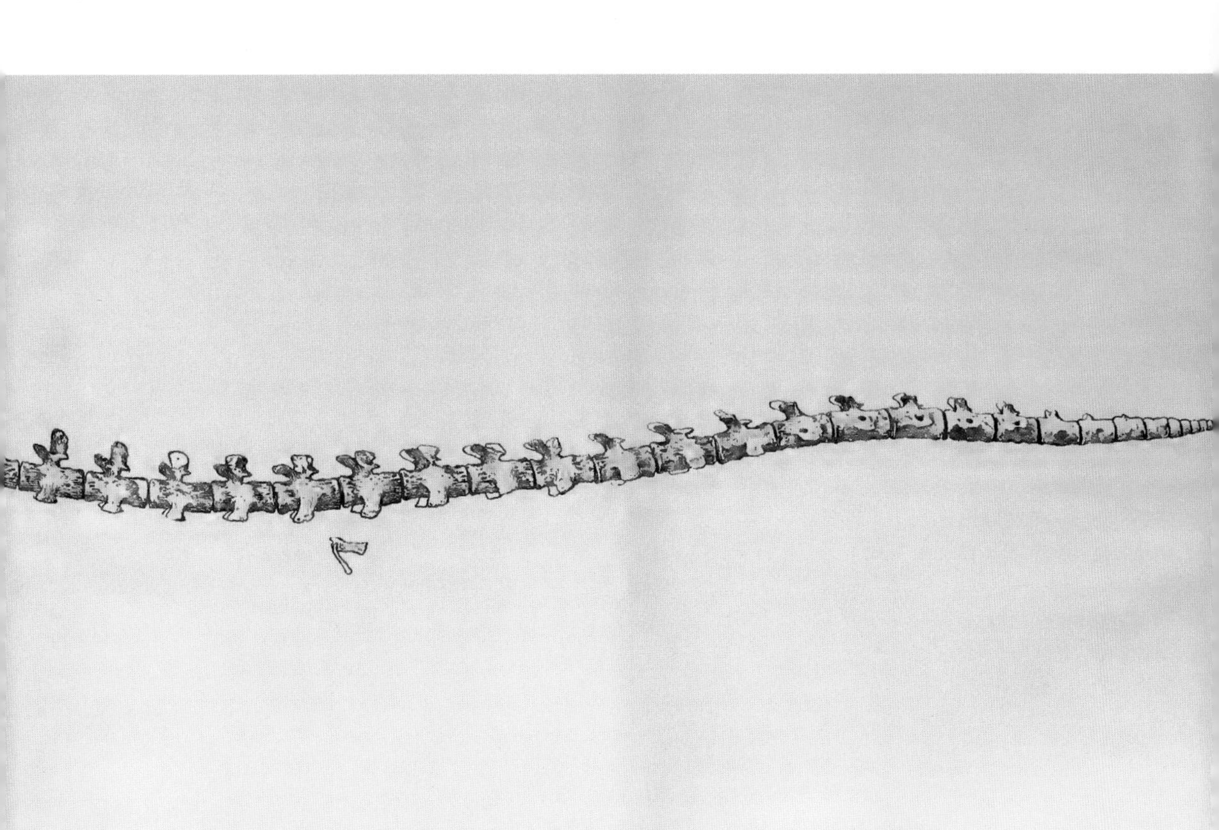

**James Williams Gidley (1866–1931), *Proceedings of the U.S. National Museum*, vol. 44, 1913; A recently mounted *Zeuglodon* (now *Basilosaurus*) skeleton.**

Sometimes confusingly called *Zeuglodon*, this is a 1913 reconstruction of a whale fossil of the genus *Basilosaurus* (see page 60). The origin of whales may be convincingly traced back to hoofed ancestors, and many ancient whales have now been unearthed in Wadi El Hitan, the "Valley of the Whales" in Egypt, and also in Pakistan. Many large forms still bore tiny hind limbs and one species, the 50-million-year-old *Pakicetus*, is effectively a dog-sized cetacean skeleton standing on four well-formed legs.

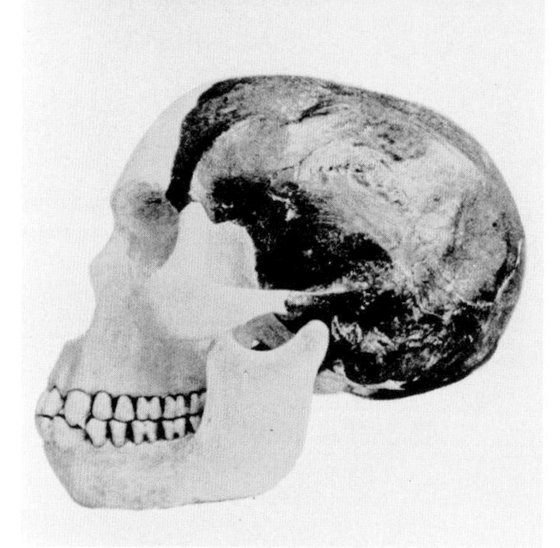

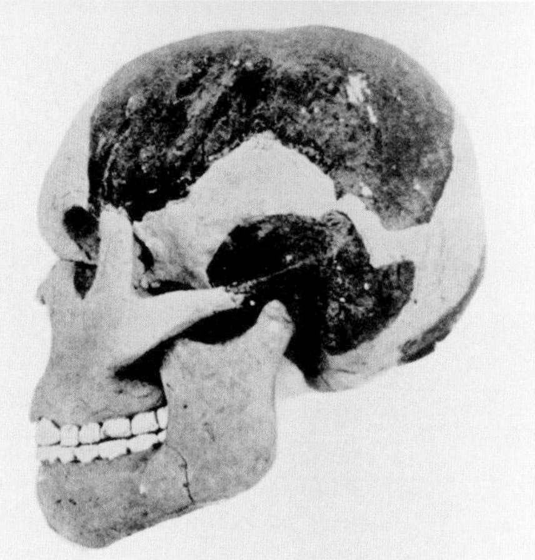

**Top: Excavations underway at Piltdown, Sussex, England; From left: Venus Hargreaves, Arthur Smith Woodward, and Charles Dawson, with Chipper the goose, circa 1908. Above: "Piltdown Man" cranium and mandible as reconstructed by Arthur Smith Woodward (left) and Arthur Keith (right), 1912.**

One of the most unedifying episodes in the history of paleontology was the "discovery" of "Piltdown Man" in Sussex, England, in 1912. Publicized as a distinctively English fossil to add to the discoveries in the Neander Valley, in North Rhine-Westphalia, and Abri de Cro-Magnon in the Dordogne, Piltdown Man seemed to possess a large brain-case but a more primitive jaw. This "first Englishman" was even claimed to have been found near to an elephant bone carved into the shape of a cricket bat. It is surprising that it took more than forty years for Piltdown Man to be exposed as a fake produced in a misguided bout of English exceptionalism—a modern human skull with a great ape's mandible artificially stained to appear similar, and with teeth rasped flat.

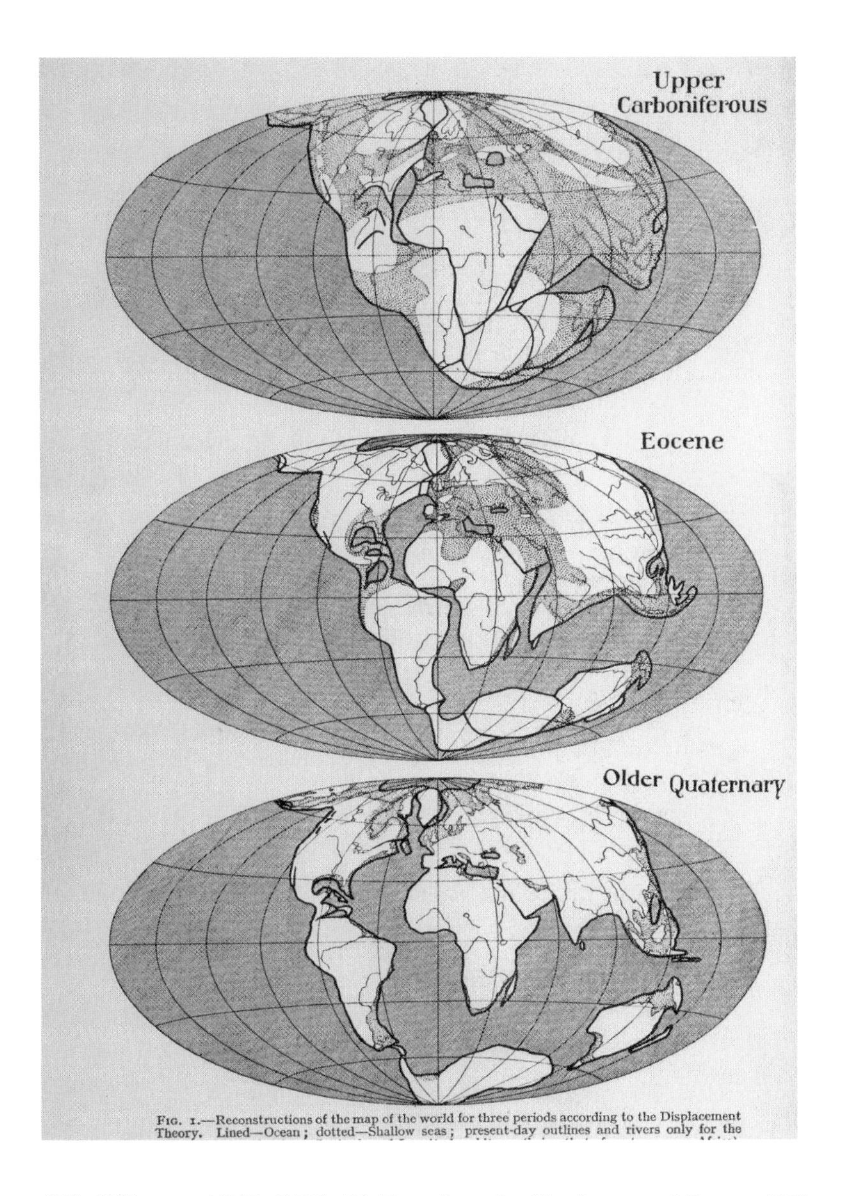


FIG. 1.—Reconstructions of the map of the world for three periods according to the Displacement Theory. Lined—Ocean; dotted—Shallow seas; present-day outlines and rivers only for the

**Alfred Wegener (1880–1930), *Die Entstehung der Kontinente und Ozeane*, 1920; Maps showing continental drift.**

Although less enthusiastically received than Piltdown Man, this other 1912 "discovery" was to prove more durable. A polymath with interests in astronomy, meteorology, and geology, Wegener would today perhaps be classed as a "planetary geologist," but his diverse interests seem to have made him something of a scientific outsider. Yet it was with an outsider's unfettered mind that Wegener drew on previous observations that the coastlines of some continents have complementary shapes, and also bear suspiciously similar fossils, to suggest that the Earth's major land masses are in continual motion over its surface. Although of central importance in paleontology, Wegener's theories were largely ignored until the 1960s, partly because he could not suggest a mechanism for his "continental drift" and partly because contemporary methods could not detect the continents' inch-per-year wanderings.

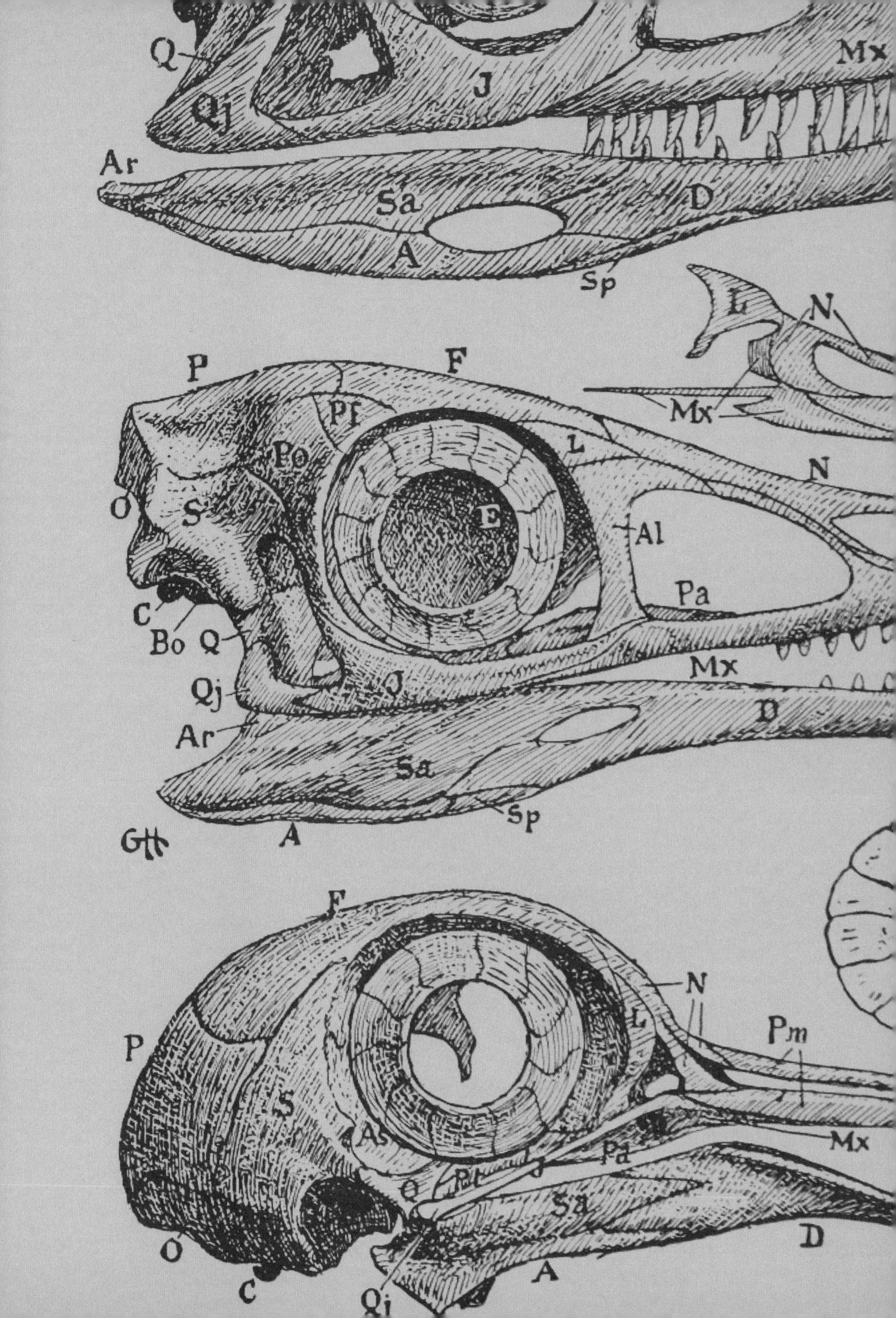

Q
Qj
J
Mx
Ar
Sa
D
A
Sp
L
N
Mx
P
F
Pf
Po
L
N
O
S
E
Al
Pa
C
Bo
Q
Qj
Mx
Ar
D
Sa
Sp
A
Gtt
F
N
P
L
Pm
S
As
Mx
Pa
Pt
Q
Sa
D
O
A
C
Qj

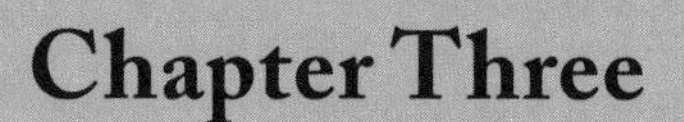

Chapter Three

# FROM INDIANA JONES TO IRIDIUM ANOMALY

1920–1980

Edwin Goodrich (1868–1946), *Studies on the Structure and Development of Vertebrates*, 1930; From top: Right view of skulls of *Euparkeria capensis*, *Archaeornis siemensi* (Berlin specimen of *Archaeopteryx*), and *Columba domestica* (modern pigeon).

# FROM INDIANA JONES TO IRIDIUM ANOMALY

## 1920–1980

In the middle years of the twentieth century, paleontology changed beyond recognition. This era begins with adventurers plundering distant plains for all the fossil treasure they could lay their hands on, and ends, by way of an economic depression and a world war, with university-led, scientifically focused paleontology as we now know it.

The driving force behind many early-twentieth-century American fossil-hunting expeditions was the great patriarch of the American Museum of Natural History in New York, Henry Fairfield Osborn (1857–1935). The son of a shipping tycoon, Osborn felt paleontology should be a hierarchical endeavor, and that he was not necessarily the person who should be grubbing around in the dirt. President of the Museum from 1906 to 1933, Osborn directed many of the excavations which have left that institution's bony coffers so remarkably full. By holding the purse-strings, Osborn acquired the right to name and publish the resulting discoveries in his own name and has left a vast scholarly legacy, especially on mammalian

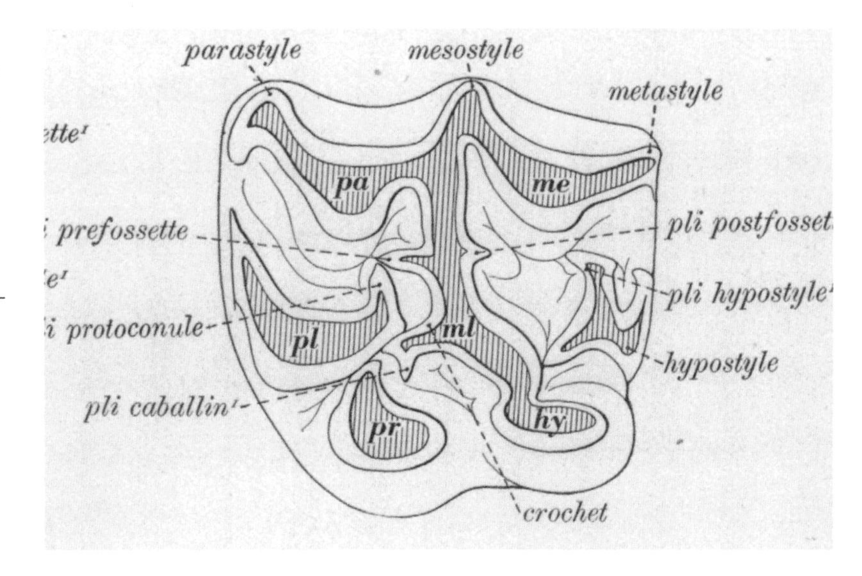


**Henry Fairfield Osborn (1857–1935), "Equidae of the Oligocene, Miocene, and Pliocene of North America," *Memoirs of the American Museum of Natural History*, vol. II, 1918; Accessory folds of the upper molars of the fossil horse *Merychippus*.**

evolution and dinosaurs—two of his most evocative name-inventions are *Velociraptor* and *Tyrannosaurus rex*. Despite his undoubted contributions, Osborn's scientific innovations were marred by his insistence on an uneasy overlap between theology and evolutionary biology, and his highly stratified (in other words racist and eugenicist) view of the populations which constitute our own species.

**Roy Chapman Andrews examining George Olsen's first find of dinosaur egg in Mongolia, in 1925.**

An occasional beneficiary of Osborn's institutional largesse was Barnum Brown (1873–1963). Born into a rural agricultural community, Brown was named after P. T. Barnum and, indeed, "Mr Bones," as he was later known, did acquire his namesake's penchant for showmanship. Brown's long career spanned the two classic eras of American paleontology—the "Bone Wars" (see page 96) and Osborn's tenure—and he could always be relied on to send railcars literally full of fossil bones back to New York. Often pictured wearing his trademark enormous fur coat, he was immensely resilient in the face of hardship, and frequently disappeared on adventures to Greece, Abyssinia, Burma, South Asia, and, of course, the great North American West. It was Brown who actually discovered *Tyrannosaurus rex*, in 1902, and he recounted with enthusiasm that fossil bones were so abundant in Wyoming that he once encountered a shepherd whose home was entirely built from them. Always a charmer, Brown was what was then called a "ladies' man," inveigled his way into working for US military intelligence in both world wars, and even acted as a consultant on Disney's *Fantasia*. He became curator at the New York museum, although the desk job clearly never suited him, and he was eventually edged out of his post in 1942.

Indeed, the Great Depression and Second World War marked the end of the romantic era of paleontology. Nations' financial agendas shifted and museums' endowments withered—the New York museum stopped funding expeditions in 1932, while London shifted its focus more toward academic research. Many organizations switched to finding fossil fuels rather than fossil animals, although Barnum Brown was able to arrange a financial stay of execution by sweet-talking expedition funding from the dinosaur-becrested Sinclair Oil company. However, the essence of paleontology had changed forever. It now became a university-led

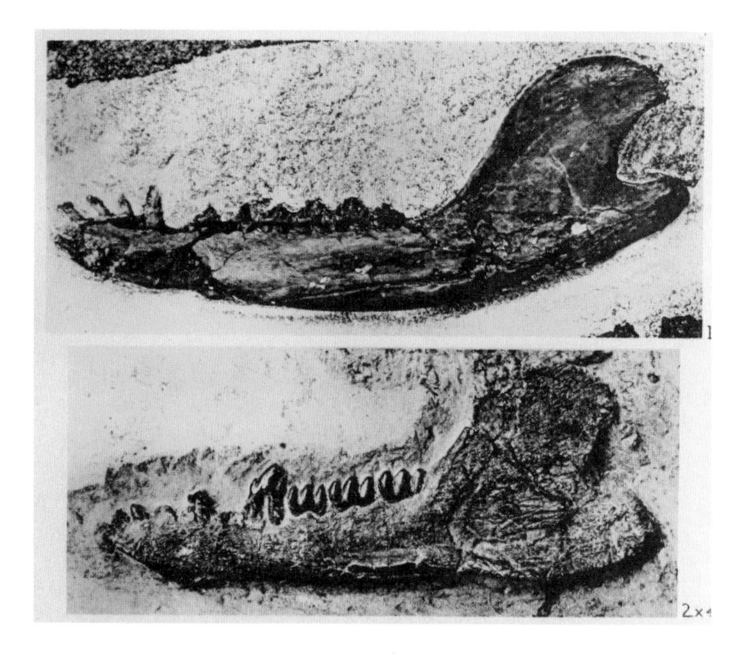

**George Gaylord Simpson (1902–1984), *A Catalogue of the Mesozoic Mammalia*, 1928; Mandibles of the extinct mammals *Phascolotherium bucklandi* (top) and *Triconodon mordax* (above).**

academic endeavor—less dashing than in the past, but better placed to properly understand the real implications of new fossil discoveries, including those churned up in the cuttings of the American post-war roadbuilding frenzy.

Once itself an offshoot of the Victorian physical sciences, after the war paleontology became ever more interdisciplinary, drawing on advances in geology, chemistry, physics, and even astronomy, as well as the less rock-bound biological sciences. A good example of this was the discovery of physico-chemical methods of dating ancient materials. In fact, radioactive decay had first been used to estimate the age of rocks in 1910, and had suggested a truly fabulous antiquity for the Earth. Later, from the 1940s, carbon-14 dating was used to date fossil animals—this heavy isotope of carbon is continually formed by the impact of cosmic rays on atmospheric nitrogen, and is then absorbed by living organisms. Once those organisms die, that carbon-14 is effectively "locked in" and degrades predictably over time into other isotopes, thus allowing the epoch of the interred creatures' death to be calculated. A similar principle applies to oxygen isotope analysis of ice cores—$H_2O$ containing oxygen isotopes with different "weights" flows through the global water cycle in subtly different ways, and can be used to determine ancient patterns of climate and glaciation.

In the 1960s, continental drift (see page 125), the decades-old suspicion that the continents creep around the planet's surface carrying their inhabitants, found a supporting mechanism in the theory of plate tectonics (see page 166). Geology also provided another way to determine the actual age of fossils: magnetostratigraphy. This technique exploits the fact that the north-south orientation of the Earth's magnetic field occasionally reverses, at rather erratic intervals counted in hundreds of thousands of years. And conveniently, these global reversals are recorded in the magnetization of contemporary sediments and volcanic lava flows, providing us with a handy "rock clock" with which to date fossils across the world. For example, the "Olduvai reversal" took place approximately 1.85 million years ago, and has been invaluable in establishing the chronology of human evolution.

The 1960s also saw a complete reorganization of how animals are classified. The 1966 publication of *Phylogenetic Systematics* by Willi Hennig (1913–1976) led to the replacement of older, arbitrary systems of animal organization by a novel, simple, almost mathematical system called "cladistics." The "clade" (Greek for "branch") became the only valid taxonomic unit, and replaced genera, classes, orders, and families. The clade was defined as all the descendants of a single ancestral species, from which those descendants have inherited particular defining, uniting characteristics. Thereafter, evolutionary trees have looked more logical, almost minimalist (see page 167), and represent scientific *theories* based on current evidence rather than subjective pigeonholing of species. And it is not just fossils that can be used as evidence to construct cladistic trees: genes can too, and conveniently enough DNA was now amenable to comparison, its structure having been elucidated in 1953 by James Watson, Francis Crick, Rosalind Franklin, and Maurice Wilkins.

**Pierre-Paul Grassé (1895–1985), *Traité de Zoologie: Agnathes et Poissons*, vol. XIII, 1958; Dorsal view of a partial reconstruction of certain internal organs of *Poraspis pompeckji* (a jawless fish).**

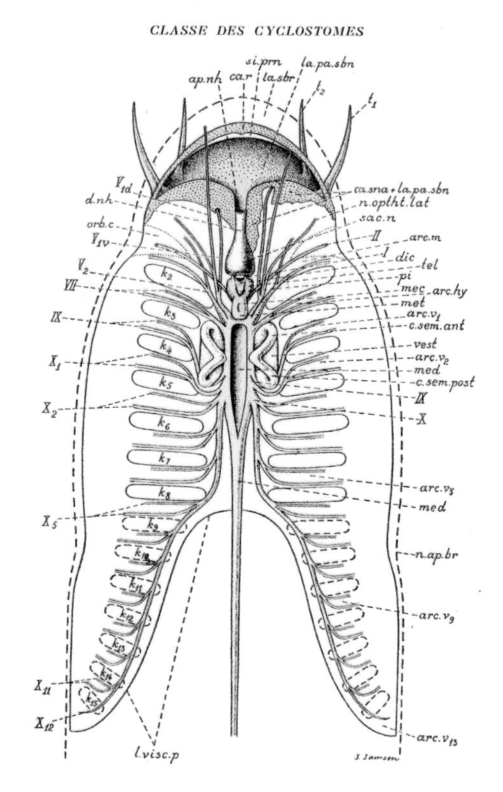


However, paleontology will always need fossil discoveries if it is to progress, even if it sometimes takes a while for the importance of those fossils to be recognized. In the 1940s, the eyes of Australian geologist Reg Sprigg (1919–1994; see page 154) fell upon some intriguing fossils in the Ediacaran Hills of South Australia, apparently while he was eating his lunch. They appeared to be animals or plants, some frond-like and some looking rather like quilts, and Sprigg realized they were unusual and probably very ancient. Despite this, he was largely ignored and the importance of what eventually became known as the Ediacaran biota (see page 154) was only gradually realized over the following decades, partly due to the definitive dating of a similar specimen from the UK. Remarkably, these organisms turned out to be the most ancient fauna/flora then known, dating from 600–540 million years ago—even older than the Burgess Shale (see page 118)—and they do not seem to correspond to any modern animal groups. Even today the Ediacaran biota still hold a unique place in the known history of life on Earth, a reminder of how a single discovery can turn our accepted ideas upside down.

In some ways, our understanding of dinosaurs came full circle in the 1960s. Originally assumed to be chilly, lethargic behemoths when they were first discovered, dinosaurs went through a more vivacious phase in the late nineteenth century (see, for example, page 108), but had since sunk back into a lumpen cold-bloodedness. However, things were to change with the reappraisal of *Deinonychus* or "terrible claw" (see page 172) by John Ostrom (1928–2005) and Robert Bakker (b.1945) at Yale University. This beast was now presented as something akin to an oversized killer-roadrunner, an intelligent, agile, warm-blooded predator with a horizontal body like an arrow in flight, bounding along on feet bearing scythe-like claws. More recent evidence from trackways and nests, and the discovery of dinosaur feathers, have extended this "Dinosaur Renaissance," confirming that dinosaurs were not the dull gray reptiles of mid-twentieth-century depictions. In fact, larger dinosaurs probably had more trouble staying cool

**Alfred Sherwood Romer (1894–1973), *Osteology of the Reptiles*, 1956; Dorsal view of a *Kotlassia* skull.**

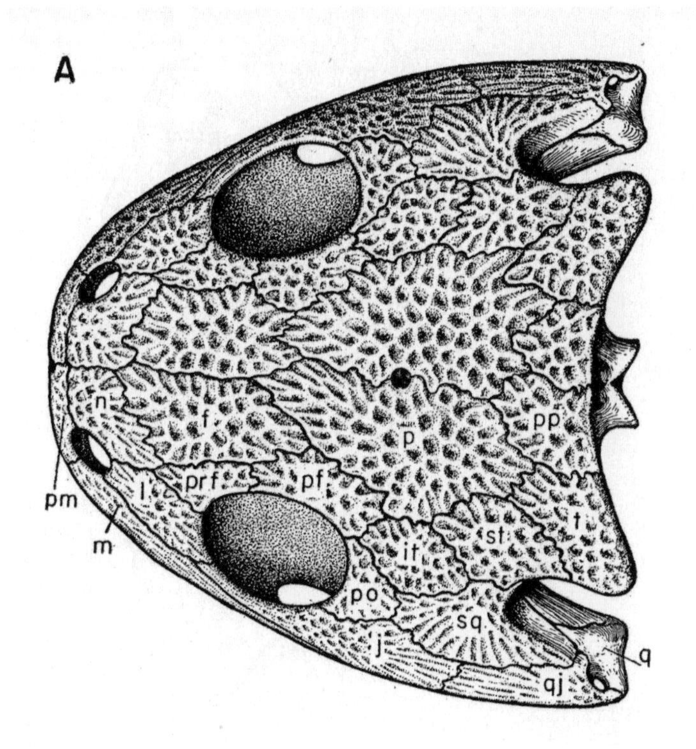

**Greater roadrunner (*Geococcyx californianus*).**

than staying warm, and smaller ones mark the origin of modern birds' diverse, high-energy, active lifestyles. A comparison of the dinosaurs in the two greatest-ever dino-movies, the "pre-renaissance" *King Kong* and the "post-renaissance" *Jurassic Park*, shows just how much our views of the terrible lizards have changed (see page 204).

Yet all good things must come to an end, and that includes the dinosaurs, or at least the dinosaurs we do not now call birds. Scientists had long wondered what caused the relatively abrupt disappearance of the non-avian dinosaurs along with a large number of their contemporary organisms, and this chapter ends with the identification of key evidence for the leading theory explaining the death of the dinosaurs. In 1978 geologists discovered vestiges of a vast crater formed by an extraterrestrial impact in the sea near modern-day Chicxulub, in Mexico, and 1980 saw the announcement of the "iridium anomaly" (see page 178). This sudden surge in the concentration of the element iridium in rocks deposited around the time of the great extinction has been claimed to confirm the impact of an enormous celestial object. Suddenly the ancient world did not seem like such an unchanging place after all—cataclysm was back.

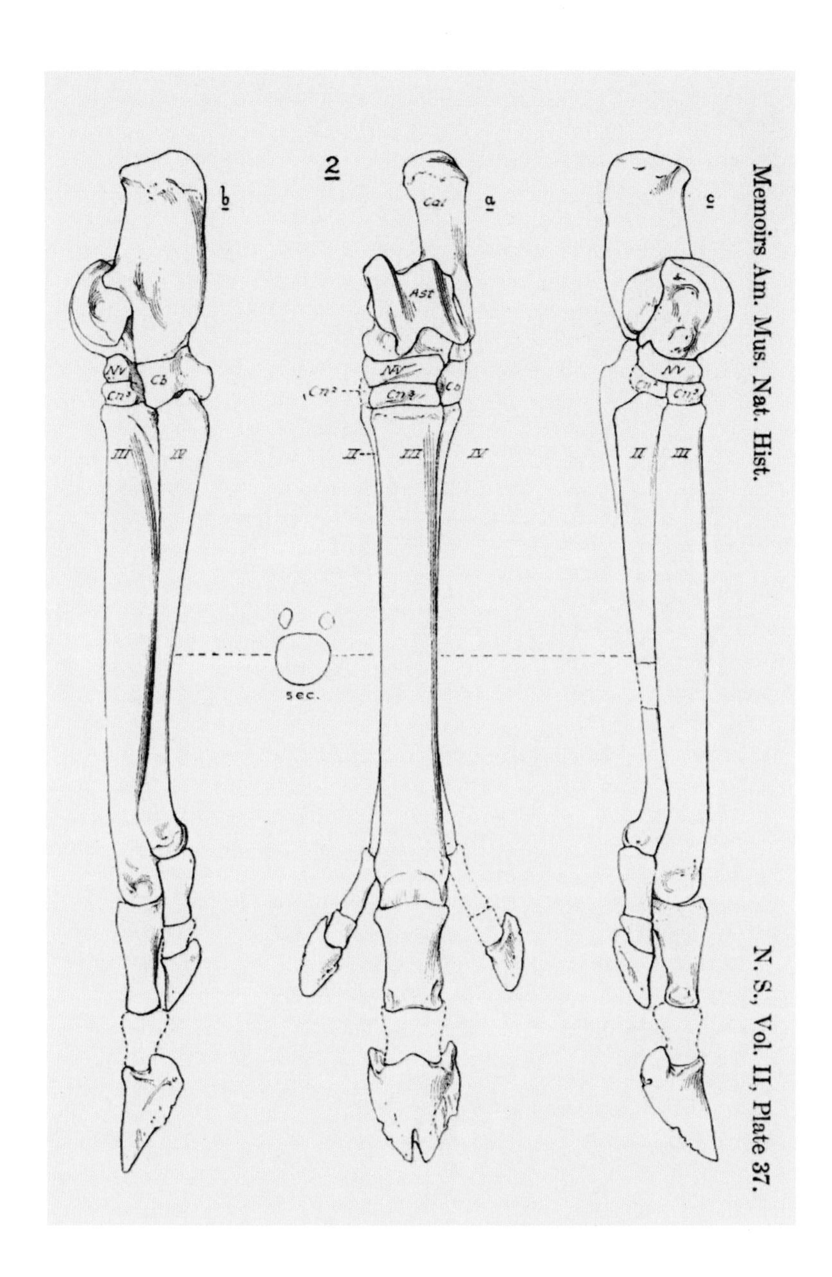


**Henry Fairfield Osborn (1857–1935), "Equidae of the Oligocene, Miocene, and Pliocene of North America,"** ***Memoirs of the American Museum of Natural History*****, vol. II, 1918; Fore and hind feet of the fossil horse** ***Parahippus*** **from Sheep Creek, Nebraska.**

The horse has one of the most complete fossil records of any species, largely thanks to Henry Osborn's expeditions to the American West. Indeed, the slow transformation from small, multiple-toed forms to today's large, single-toed equids became a common trope for biology textbooks.

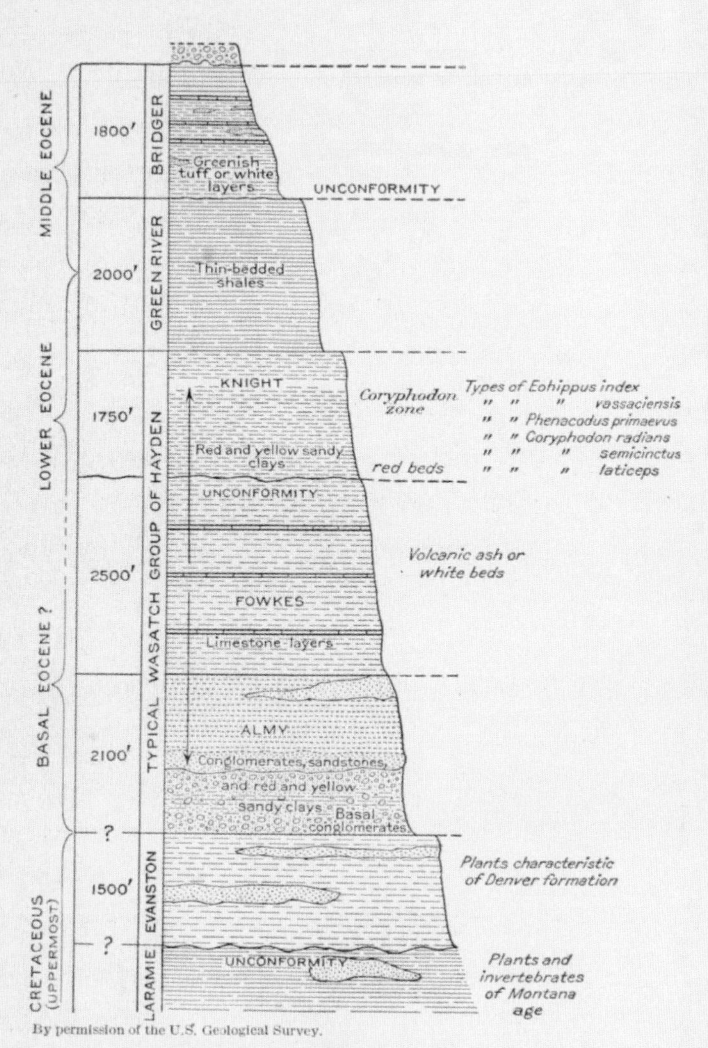


By permission of the U.S. Geological Survey.

Fig. 36. — Scale section of the Lower and Middle Eocene of southwestern Wyoming, showing the relations of the "typical Wasatch group" of Hayden (A); Modified from Veatch, 1907.

**Henry Fairfield Osborn (1857–1935), *The Age of Mammals in Europe, Asia, and North America*, 1910; Scale section of the Lower and Middle Eocene of Southwestern Wyoming.**

Many of Osborn's works contain detailed stratigraphic surveys of the areas in which fossils were discovered, in this case the Eocene strata of Wyoming. The Eocene Epoch occurred between 56 and 34 million years ago, an era of diversification that saw the spread of mammals after the demise of the dinosaurs 66 million years ago.

**Skull and jaws of *Velociraptor mongoliensis* found by Henry Fairfield Osborn, now in The American Museum of Natural History.**

Later made famous by its starring role in the *Jurassic Park* movies (see page 204), the genus *Velociraptor* (fast-grabber) was characterized early in the twentieth century. *Velociraptors* were actually quite small, less than 2ft (60cm) tall, weighed as much as a collie dog, and may have been covered in feathers. They did, however, share with their larger relatives a single, oversized, sickle-shaped claw on each of their feet, presumably adapted for slashing at their prey.

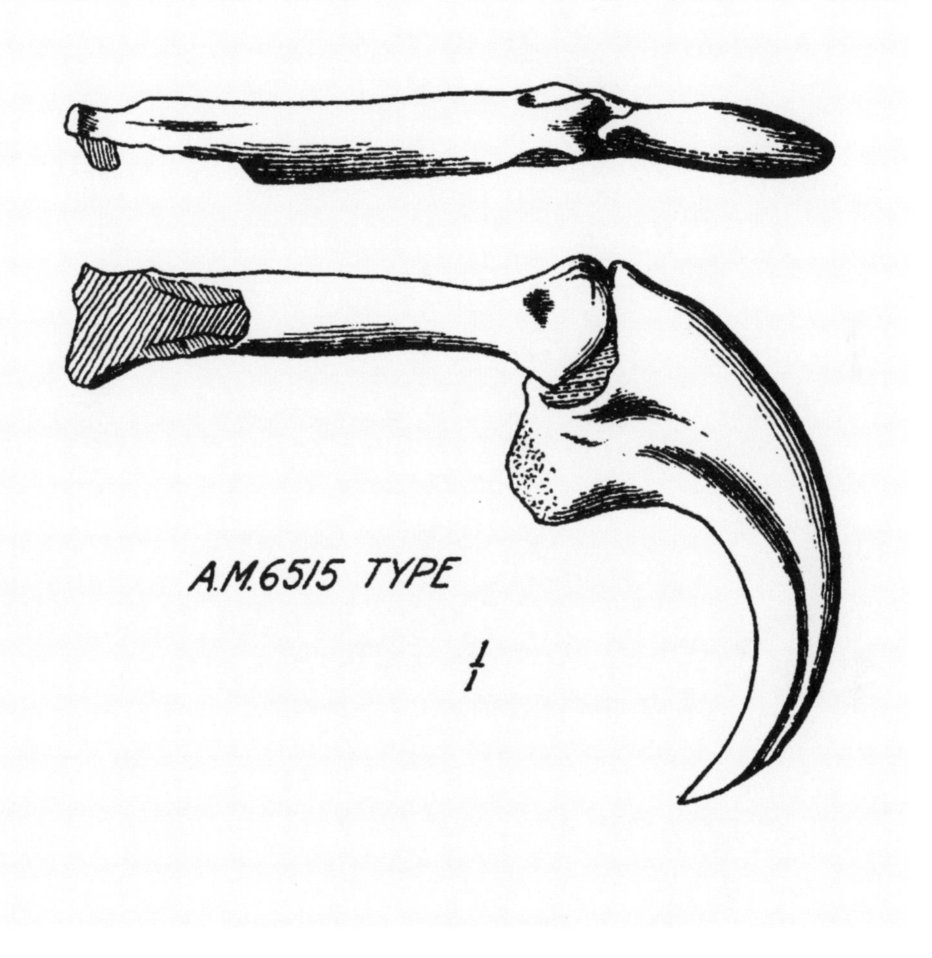


**Henry Fairfield Osborn (1857–1935), "Three New Theropoda, Protoceratops Zone, Central Mongolia,"** ***American Museum Noviciates*****, vol. 44, 1924, The American Museum of Natural History; Phalanges (finger bones) of** ***Velociraptor mongoliensis*****.**

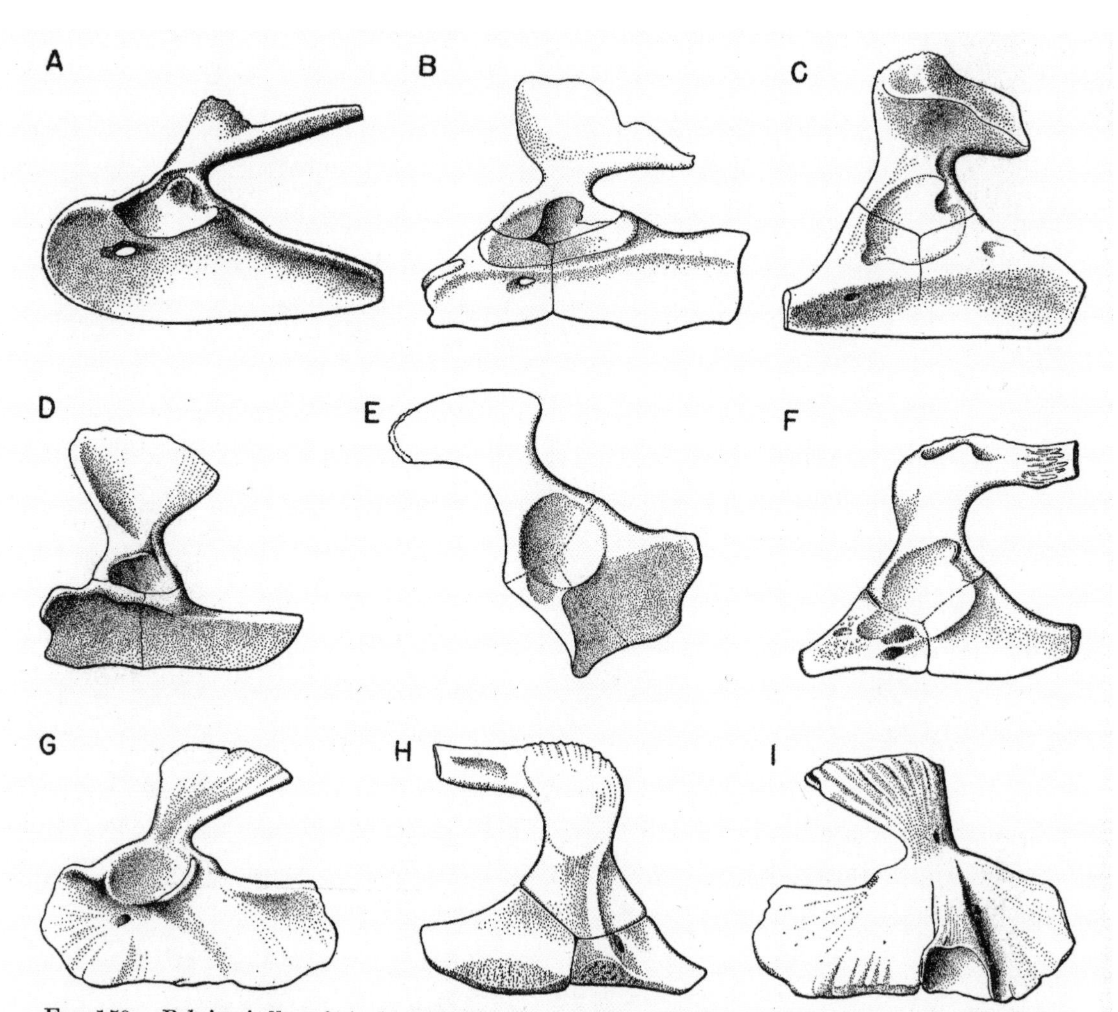


FIG. 150.—Pelvic girdles of *A*, the embolomere *Archeria*; *B*, *Seymouria*; *C*, *Diadectes*; *D*, *Nyctiphruretus*; *E*, *Embrithosaurus*; *F*, *Limnoscelis*; *G*, *Labidosaurus*; *H*, *I*, internal views of left girdle of *Limnoscelis* and *Labidosaurus*. (*B*, after White; *E*, after Broom; *G*, *I*, after Broili.)

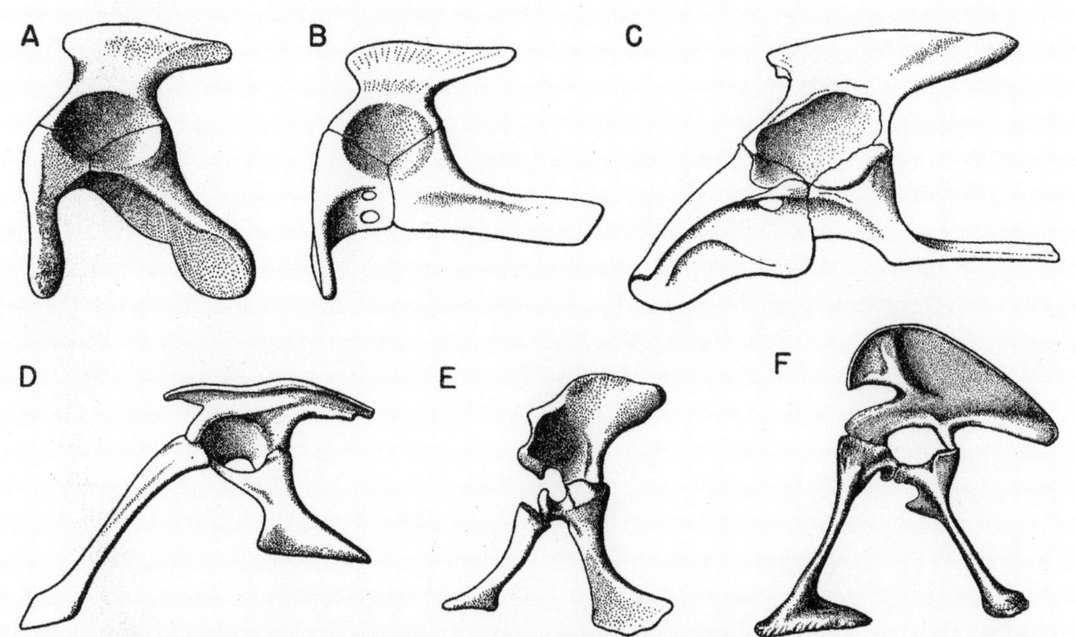


FIG. 155.—Pelvic girdles of archosaurs. *A*, ***Erythrosuchus***; *B*, ***Euparkeria***; *C*, ***Machaeroprosopus***; *D*, ***Protosuchus***; *E*, ***Alligator***; *F*, ***Antrodemus***. (*A*, after Huene; *B*, after Broom; *C*, after Camp; *D*, after Colbert; *F*, after Gilmore.)

**Alfred Sherwood Romer (1894–1973), *Osteology of the Reptiles*, 1956; Pelvic girdles of a variety of basal reptiles (opposite) and archosaurs (above).**

Romer was one of the most influential paleontologists and vertebrate morphologists of the twentieth century, publishing many popular books as well as copious basic research. He is particularly known for his work on the skeletons of land vertebrates, especially reptiles. The tetrapod pelvis usually comprises three main components, the ilium, ischium, and pubis, joined at the hip socket, and these images trace its variations among different reptiles, extinct and living.

# ROY CHAPMAN ANDREWS
## (1884–1960)

### The Ends of the Earth

The early twentieth century saw a series of swashbuckling adventurers scouring the American West, and sometimes the whole world for paleontological plunder. And the bucklers were never more swashed than by Roy Chapman Andrews, fossil hunter, popular author of *Ends of the Earth* and *Secrets from the Rocks*, and, most of all, adventurer.

Later the reinvigorating director of the American Museum of Natural History in New York, Andrews started his career by sweeping its floors in 1906. However, by 1920 he had pioneering surveys of the fauna of Southeast Asia, Yunnan, and the Arctic under his belt, and was now to embark on a decade-long series of famous adventures.

**Roy Chapman Andrews (1884–1960) and Henry Fairfield Osborn (1857–1935), *On the Trail of Ancient Man: A Narrative of the Field Work of the Central Asiatic Expeditions*, 1926; *Protoceratops* skull found in Shabarahk Usu, Mongolia, in 1925.**

George Olsen (left) showing Roy Chapman Andrews the nest of dinosaur eggs that he discovered in the Valley of Shabarahk Usu, Mongolia, in 1923. Photograph James B. Shackleford, 1923.

Arriving in Peking, he drove west in that new-fangled contraption, the motor car, to the Gobi Desert and Mongolia where troves of novel dinosaur and early mammal bones awaited him. His team's most famous discovery came in 1923, with clutches of eggs thought to have been laid by the "frilled" dinosaur *Protoceratops*, but now attributed to the bipedal dinosaur *Oviraptor*.

Never one to shy away from risk or self-publicity, Andrews was shot in the leg by a stray bullet while hunting and boasted: "I can remember just ten times when I had really narrow escapes from death." Indeed, it has been claimed that Andrews, with the distinctive hat he wore in many photographs, inspired the character of Indiana Jones. More likely, the influence was indirect, as Andrews' fame meant he became something of an archetype for the American adventurer in mid-twentieth-century cinema.

**Charles Gilmore (1874–1945) of the Smithsonian Institution with vertebrae from the dinosaur *Diplodocus*, *Washington Post*, September 24, 1924.**

Looking suitably forbidding in this newspaper image, Professor Charles Gilmore was a paleontologist at what is now the National Museum of Natural History in Washington D.C. He unearthed, characterized, and named many species, dinosaurs especially, and in this picture appears with four vertebrae from the large Jurassic sauropod dinosaur *Diplodocus*. This species was relatively gracile compared to related groups and, despite being 80ft (24m) long, weighed as "little" as 12 tons. The tycoon Andrew Carnegie arranged for several casts to be made of *Diplodocus carnegii* and donated to museums around the world, and ironically one such replica, London's Natural History Museum's "Dippy," is now more famous than any of its "real" exhibits.

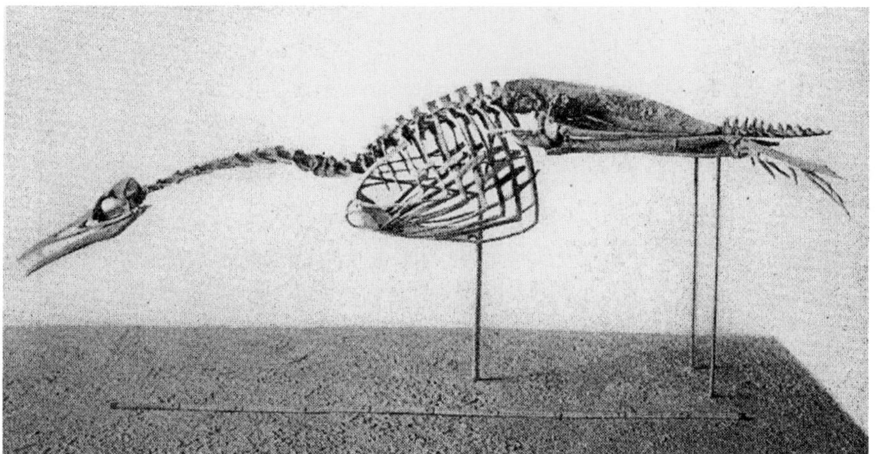

**Richard Swann Lull (1867–1957), *Organic Evolution*, 1917; Skeletons of *Hesperornis regalis* and *Hesperornis crassipes*, a toothed diving bird from the Cretaceous Period of Kansas.**

Overshadowed by the fame of *Archaeopteryx*, *Hesperornis* gives another fascinating insight into the evolution and habits of birds during the time when their non-avian dinosaur cousins still walked the Earth. This toothed, almost wingless bird has been well studied since its discovery and the consensus is that it was a marine diver, with a posture similar to modern penguins. *Hesperornis*'s fine teeth presumably helped it seize its prey, although all modern birds are toothless, presumably to reduce the weight of the jaws and head.

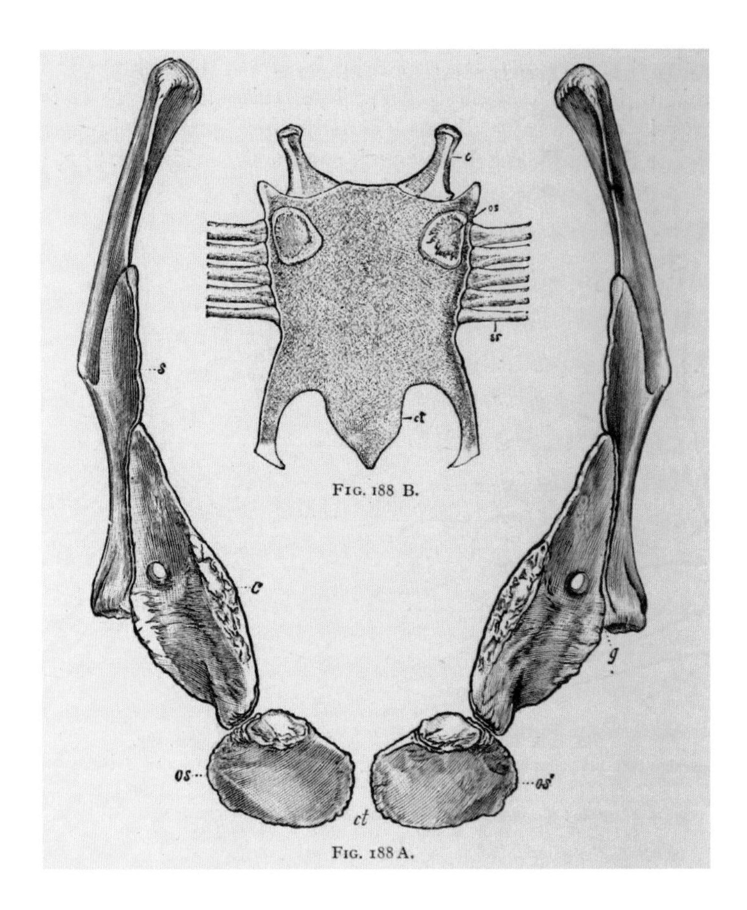


**Above: Edwin Goodrich (1868–1946), *Studies on the Structure and Development of Vertebrates*, 1930; *Brontosaurus excelsus* (Jurassic Period, Wyoming), showing the anterior aspect of the pectoral arch and presented with the breastbone of a young ostrich (*Struthio camelus*) for comparison.**

Edwin Goodrich was an English morphologist who sought to synthesize anatomy, embryology, and development into a unified scheme of vertebrate structure. His work could be argued to represent the transition point between "classical" descriptive investigations of animal structure and development, and "modern" experimental and eventually molecular studies. The image opposite is a comparison of four species of archosaur—the lineage of reptiles including pterosaurs, dinosaurs, and birds. The top two are relatively ancient forms, the third is *Archaeopteryx*, and the fourth is the domestic pigeon. Above are the breastbones and forelimb girdles (shoulder blade, collar bone, and others) of *Brontosaurus* (outer) and an ostrich (center).

**Opposite: Edwin Goodrich (1868–1946), *Studies on the Structure and Development of Vertebrates*, 1930; Right view of the skulls of (A) *Aetosaurus ferratus*, (B) *Euparkeria capensis*, (C) *Archaeornis siemensi* (Berlin specimen of *Archaeopteryx*), and (D) *Columba domestica* (modern pigeon).**

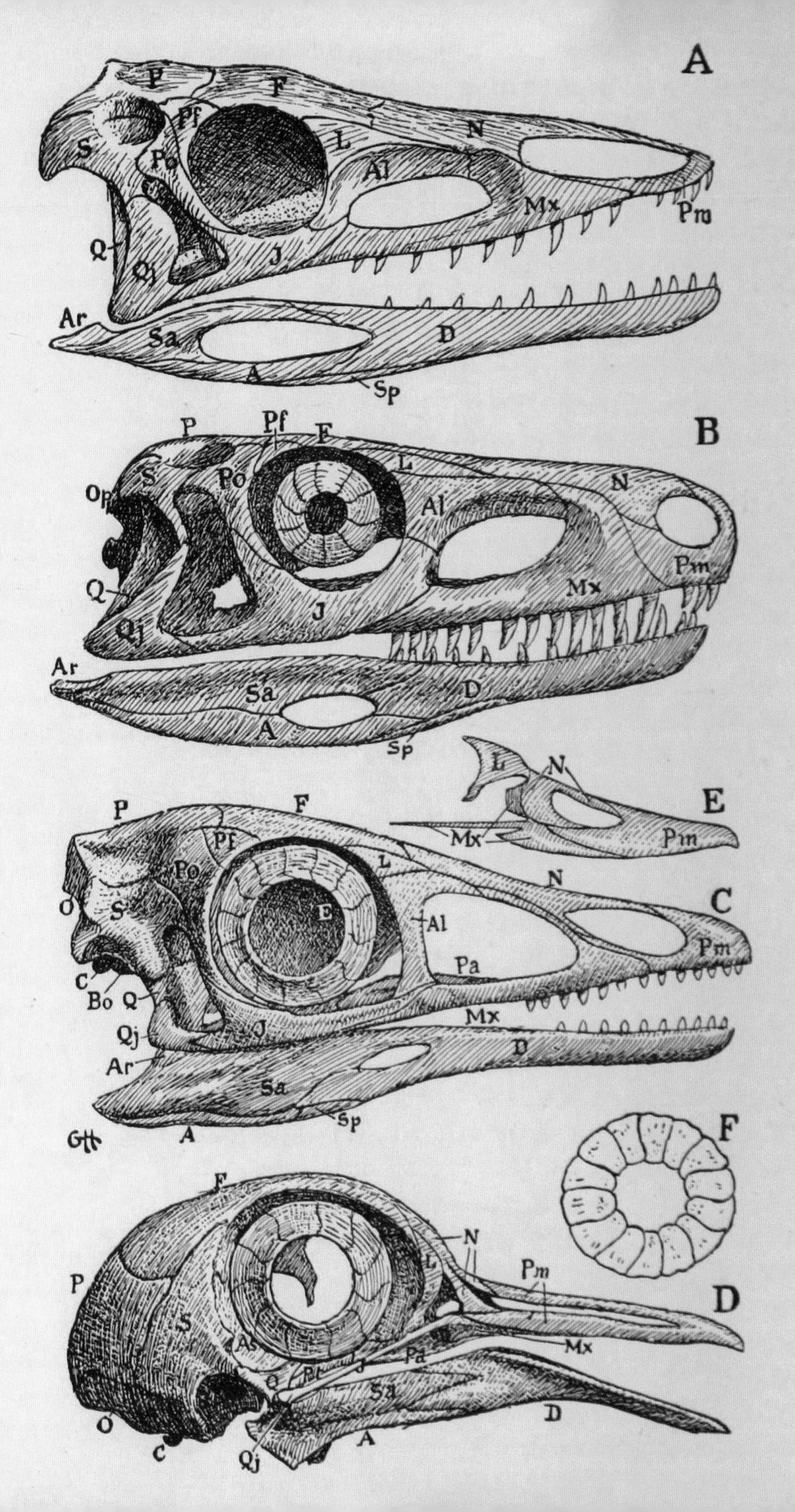
A
P
F
Pf
S
Po
L
N
Al
Mx
Pm
Q
Qj
J
Ar
Sa
D
A
Sp
B
Op
C
E
Pa
Bo
O
As
Pt
GH

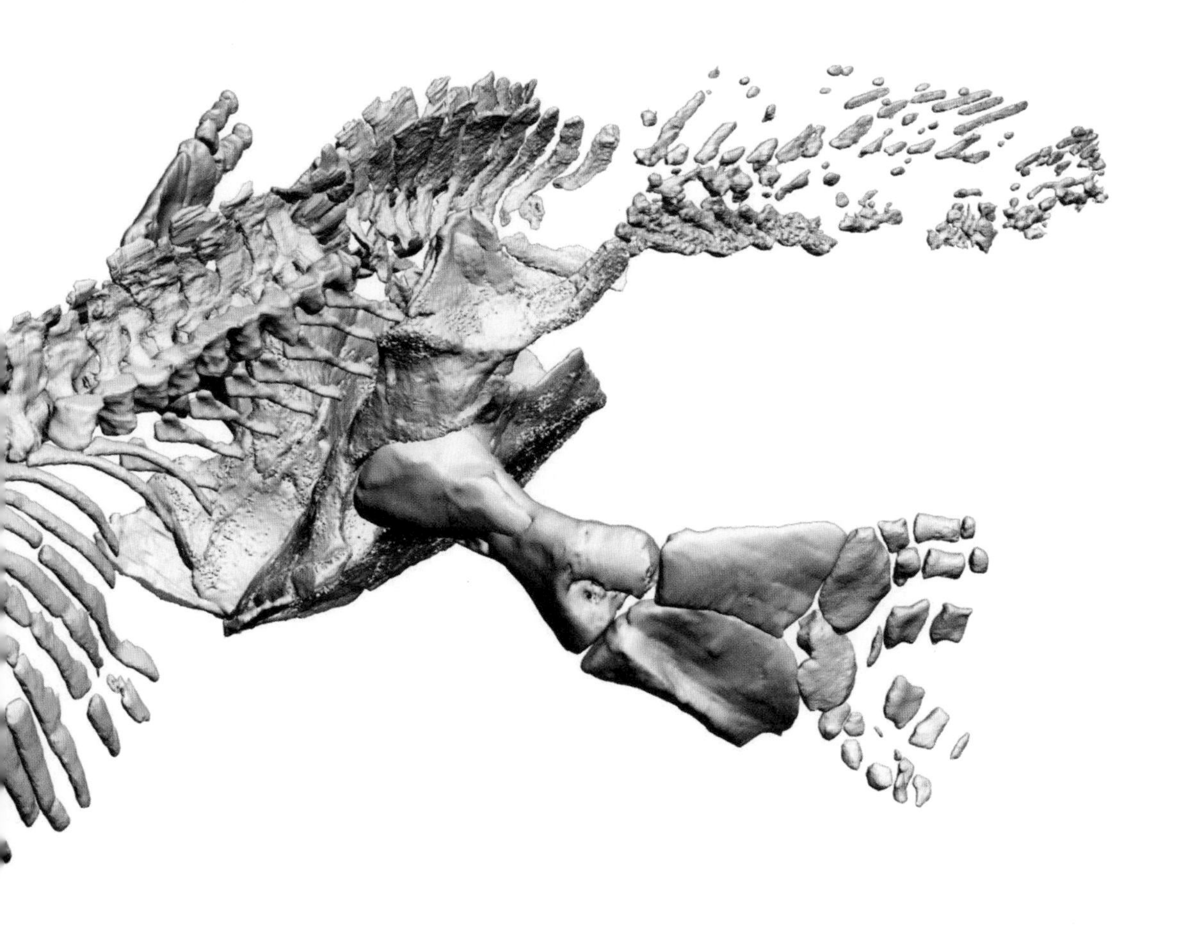

**Reconstruction by Stephanie Pierce and others, "Three-dimensional limb joint mobility in the early tetrapod *Ichthyostega*," *Nature*, vol. 486, 2012.**

Discovered by Gunnar Säve-Söderbergh in Greenland in 1932, *Ichthyostega* dates from approximately 360 million years ago, and was the first fossil investigated as a possible "link" between fish and land vertebrates. It possessed jointed limbs as well as toes—probably more than five on some of its feet—but also an extensive tail-fin supported by fish-like bony rays. In many older representations, *Ichthyostega* is depicted something like a giant salamander perched on a rock, with its tail dipped suggestively in the water of a primordial pond. Although *Ichthyostega* probably did emerge onto land at times, the paper from which this image is taken suggests that it could not walk like modern amphibians.

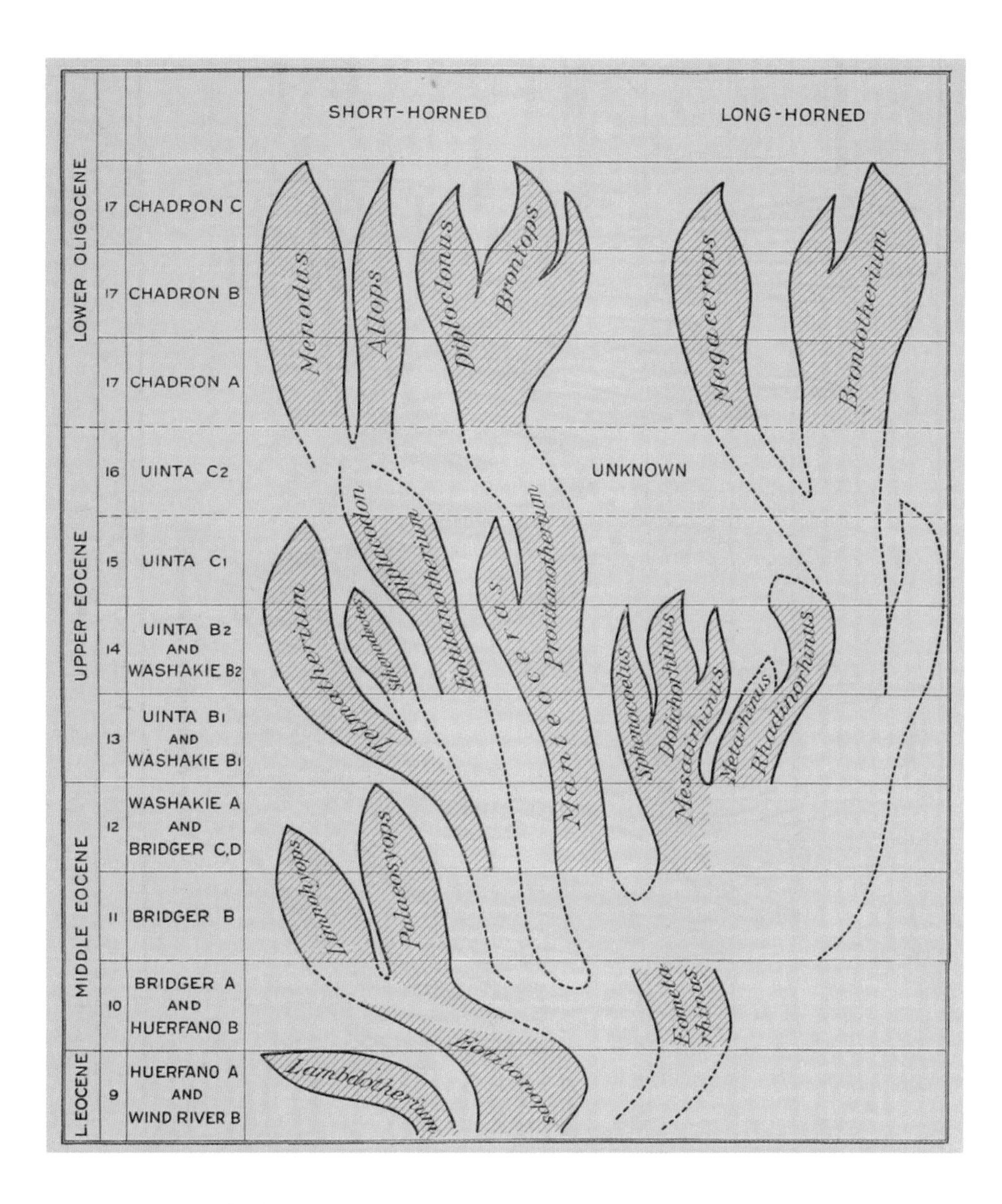


**Henry Fairfield Osborn (1857–1935), *The Titanotheres of Ancient Wyoming, Dakota, and Nebraska*, vol. 55, 1929; Evolutionary tree of the titanotheres.**

As their name suggests, the titanotheres were a group of huge mammals, and they lived across the Northern Hemisphere from 55–35 million years ago (see page 9). This titanothere family tree is based on their fairly complete fossil record in the great plains of the U.S. Three decades later, diagrams like this were to be superseded by cladograms, objective scientific hypotheses that make fewer assumptions about the evolutionary process (see page 167).

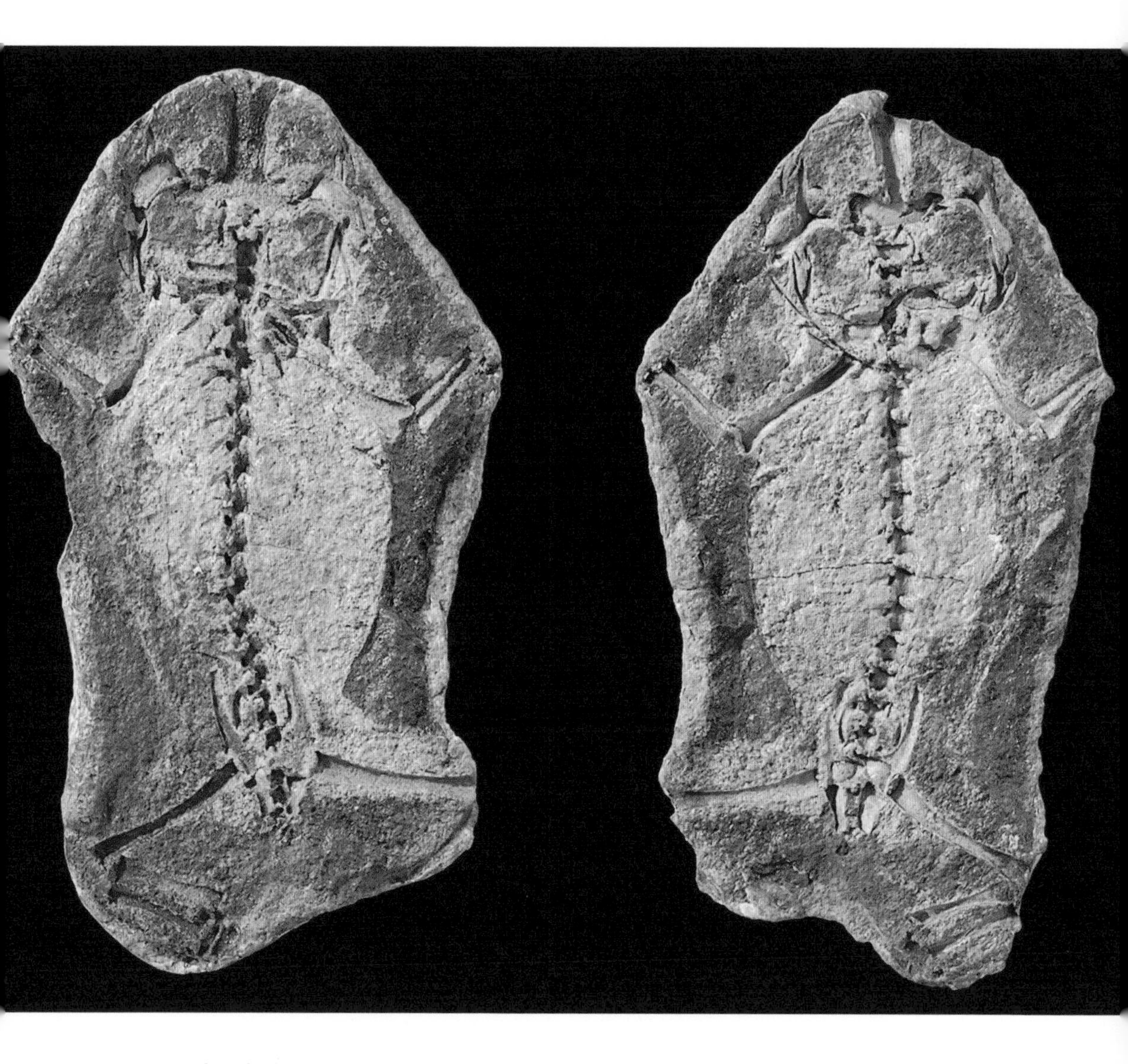

**Eduardo Ascarrunz and others, *"Triadobatrachus massinoti,* the earliest known lissamphibian (Vertebrata: Tetrapoda) re-examined by μCT scan, and the evolution of trunk length in batrachians"; *Contributions to Zoology,* vol. 85, 2016.**

Modern amphibians (or lissamphibians: "smooth amphibians") have diversified into three basic body plans—frogs and toads, salamanders and newts, and the legless caecilians. Among these, the frogs represent a striking and unusual morphology which has been successful for approximately 250 million years. Frogs seem to appear abruptly in what is a fragmentary fossil record with many of their distinctive features already present in *Triadobatrachus* (pictured here). This species already possesses skull and limbs like those of modern frogs and toads, although the spine remains long and sinuous, unlike today's truncated, compact species.

**Charles Whitney Gilmore (1874–1945), "A mounted skeleton of *Dimetrodon gigas* in the United States National Museum, with notes on the skeletal anatomy," *Proceedings of the United States National Museum*, vol. 56, 1919.**

Although *Dimetrodon*, with its distinctive sail, frequently appears in popular dinosaur books, it is, in fact, more closely related to mammals—it sits somewhere near the base of the evolutionary branch eventually leading to today's furred, live-bearing, lactating forms. Indeed, dating from 300–270 million years ago, its name means "two sizes of teeth," and differentiation of a variety of tooth types was probably crucial for mammals' eventual success. The sail remains enigmatic, however, and suggested functions have included heat dissipation, courtship, threat, spinal stability, and hearing.

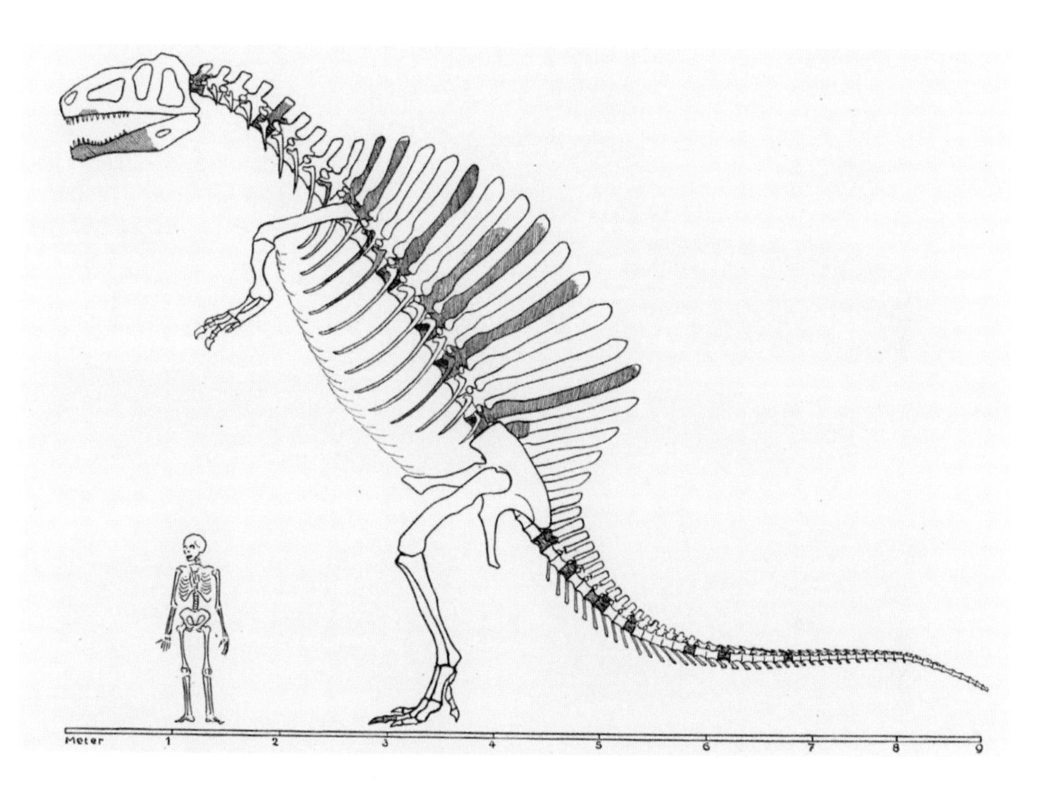


**Ernst Stromer (1871–1952), Sketch on a glass slide of *Spinosaurus aegyptiacus*, from *Abhandlungen der Königlichen Bayerischen Akademie der Wissenschaften*, 1936, vol. 33, p.65, fig. 8.**

Another ancient sail-backed animal, *Spinosaurus* was discovered in 1912 in Egypt in strata laid down approximately 95 million years ago. A dinosaur unlike *Dimetrodon*, this species bore large spines on its vertebrae which are assumed to have supported a large sail. Not long after this reconstruction was inked onto a glass slide, the first *Spinosaurus* specimen was destroyed in the Allied bombing of Munich. More recent fossils have, however, confirmed that *Spinosaurus* was a rather bizarre carnivorous dinosaur, as large as *Tyrannosaurus*, but with strangely stumpy hind limbs and possibly webbed toes, suggesting a quadrupedal, semi-aquatic habit.

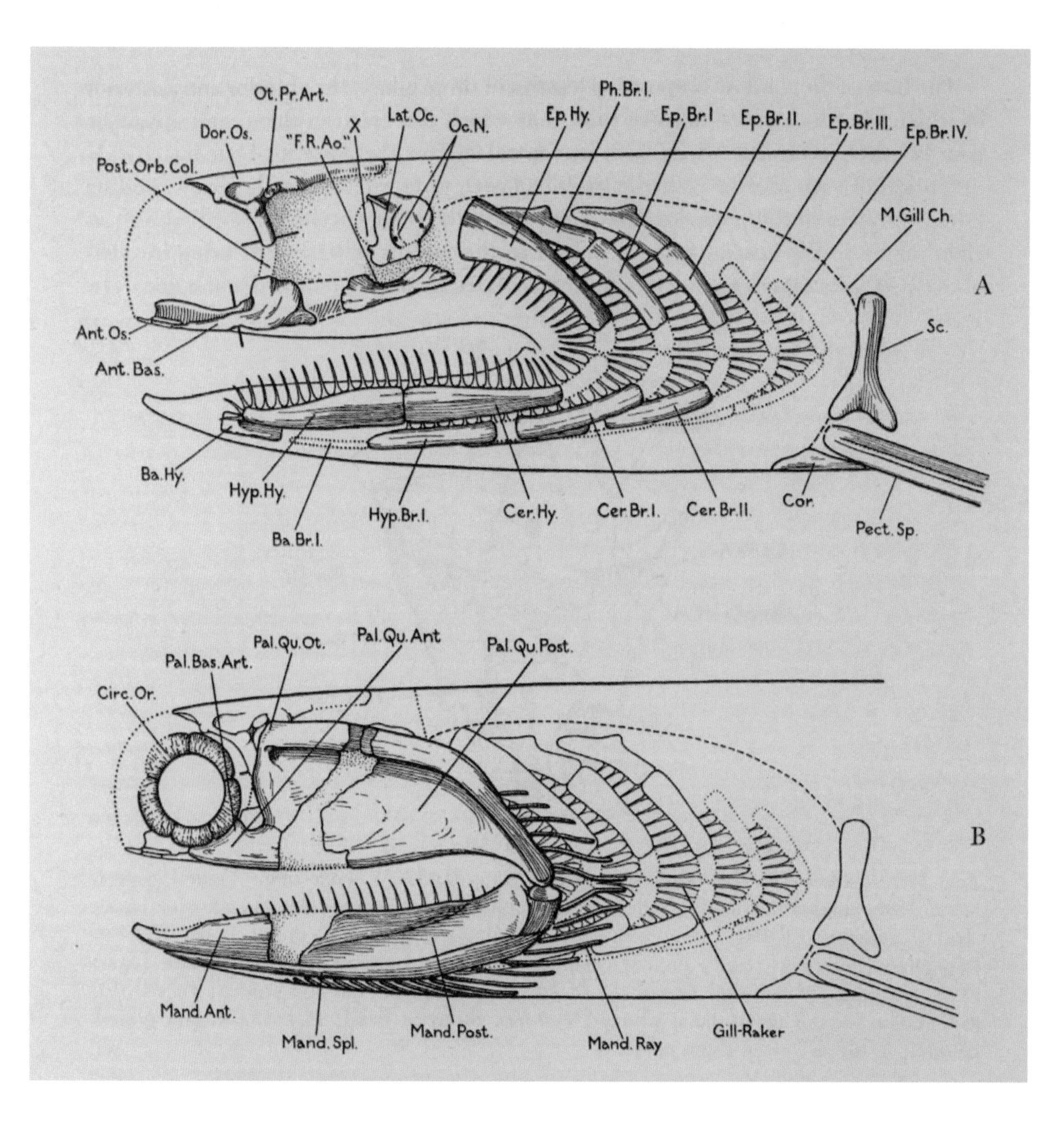


David Watson (1886–1973), "On the Acanthodian Fishes," *Philosophical Transactions of the Royal Society of London*, Series B, vol. 228, 1937; *Acanthodes*.

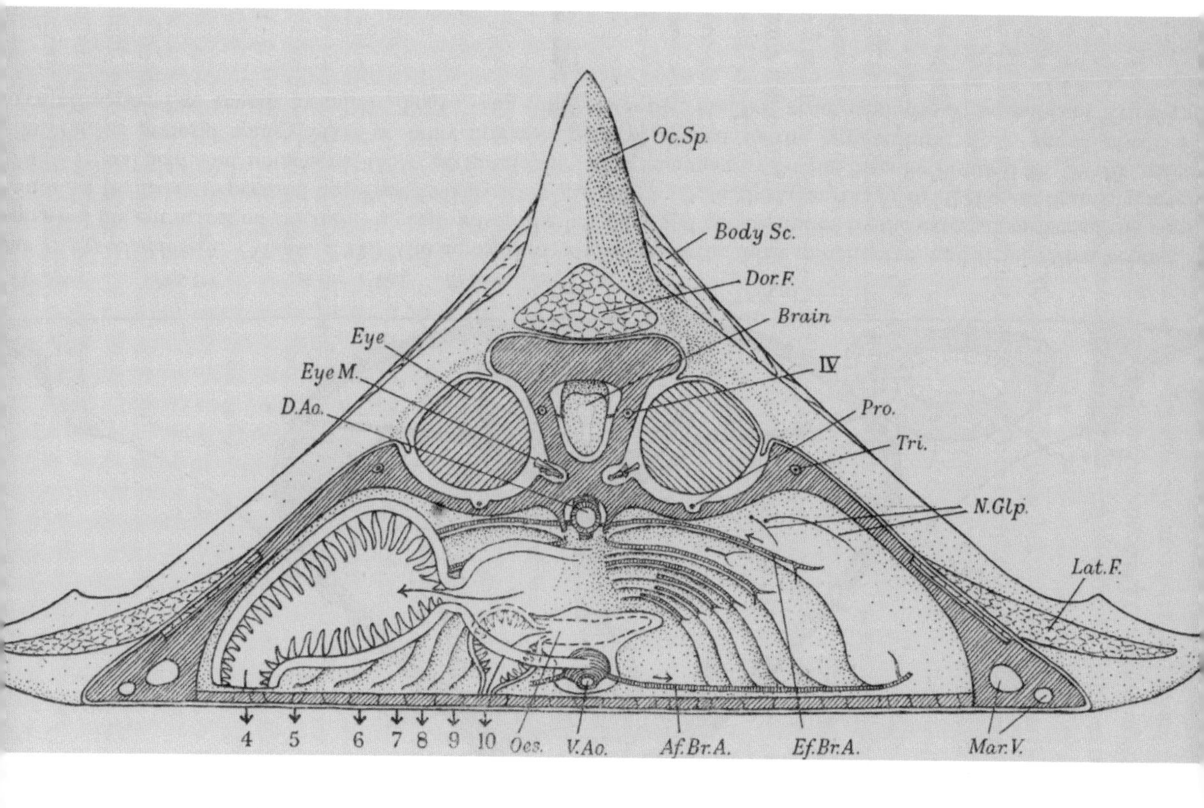


**David Watson (1886–1973), "A Consideration of Ostracoderms,"** ***Philosophical Transactions of the Royal Society of London,*** **Series B, vol. 238, 1954; Reconstruction of a transverse section through the head of** ***Cephalaspis hoeli.***

Watson was professor of zoology and comparative anatomy at University College, London, and wrote extensively on ancient fish. The images on the opposite page are of *Acanthodes*, a genus from approximately 300 million years ago, although similar genus may date back considerably further. *Acanthodes* is a confusing fossil as it mixes characteristics of cartilaginous fish (sharks and rays) and bony fish (most other fish)—and thus may lie somewhere near this fundamental bifurcation in the evolutionary story of jawed vertebrates. In contrast, the alien-like image on this page is a proposed cross section through the head of *Cephalaspis*, an armored jawless fish from 400 million years ago. It was first described by Louis Agassiz in 1843 (see page 51), and is thought to be a relative of today's un-armored parasitic lampreys.

# REG SPRIGG (1919–1994) AND TREVOR FORD (1925–2017)

## Life Before Ancient Life

The story of the discovery of the Ediacaran biota is the paleontological equivalent of discovering the ruins of the world's oldest city. In the nineteenth century, the fossil record was believed to hit a sudden barrier beyond rocks 541 million years old—the familiar segmented trilobites and shelled brachiopods simply seem to stop below the oldest Cambrian rocks. Indeed, the Cambrian is the earliest period within what was called, for this reason, the Paleozoic (ancient-life) era.

There had been suggestions of Pre-Cambrian life through the decades, but they were not taken seriously until the discovery of a strikingly novel array of fossils in the Ediacara Hills of South Australia in the 1940s by the geologist Reg Sprigg, and the precise dating in 1958 of a similar fossil discovered by the speleologist Trevor Ford in Charnwood Forest in England. Since that time similarly ancient and unusual fossils have been found across the globe, including the very oldest at Mistaken Point in Newfoundland.

**Fossil of the tree fern *Dickinsonia costata*, from the Pre-Cambrian Ediacaran biota.**

Life in the Ediacaran Sea, diorama by Terry Chase Studios, 1980, formerly displayed in the Smithsonian Institution, Washington D.C., now in the Paleontological Research Institution, Ithaca, New York.

The fossils are so hard to characterize that they are now called the "Ediacaran biota," as opposed to "fauna" or "flora," to avoid assumptions about them being plants or animals. They date from as early as 575 million years ago and are particularly difficult to assign to modern groups, although some have been claimed to be similar to corals, worms, and cuticle-less arthropods (the group including insects and crustaceans). They are often classified according to shape rather than any claimed evolutionary relationships—as filaments, tubes, fronds, discs, toroids (donuts), "quilts," and a few forms which are seemingly more complex.

The current consensus is that many Ediacaran organisms are, in fact, animals, but do not correspond to modern groups, since those groups may not yet have existed at that time. No one knows why the Ediacaran organisms became extinct, nor why complex life appears so late in the fossil record in the first place. There is reason to believe that complex life could not flourish until the Earth's atmospheric oxygen concentrations rose, but this probably happened around 1,000 million years ago. Thus, in the sparsely populated landscape of the Pre-Cambrian, the Ediacaran biota are actually a late, exceptional, and mysterious outpost.

**The Ediacaran fossil *Charnia masoni*, found in the 1950s in Charnwood Forest, Leicestershire, England, by school students Tina Negus and Roger Mason, and dated by speleologist Trevor Ford in 1958.**

**Artist's reconstruction of *Tullimonstrum* or the "Tully Monster."**

Discovered in Illinois in 1958, there can be few fossils as bizarre as *Tullimonstrum*. Extracted from 307-million-year-old strata, the creature is so strange that only recently has there been anything approaching consensus about what sort of animal it actually is—suggestions have included a finned segmented worm, a swimming slug, or a deviant squid. *Tullimonstrum* was a spindle-shaped creature with upper and lower fins, protuberant eyes held out to the side on a rigid bar, and a long, jointed snout bearing a toothed pincer or jaw. Although chemical analysis of metals in its fossilized eyes suggested similarities to squid, structural analyses have demonstrated the presence of paired gill holes and a longitudinal stiffening bar. This bar is called a notochord and is the defining feature of "chordates," the group which includes vertebrates. For this reason, *Tullimonstrum* is now thought to be another morphologically exceptional relative of modern lampreys.

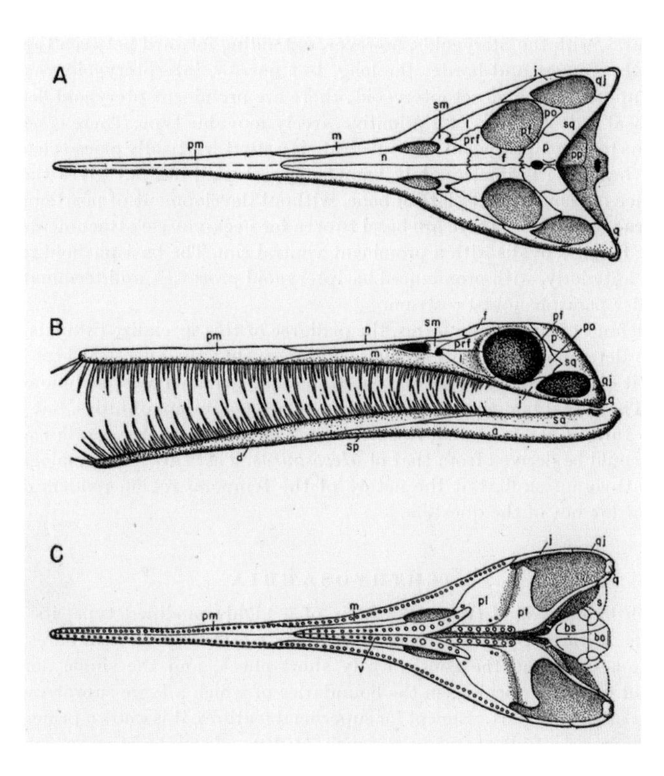


**Alfred Sherwood Romer (1894–1973), *Osteology of the Reptiles*, 1956; Dorsal, lateral, and ventral views of the skull of the early marine reptile *Mesosaurus*.**

Although the move from water to land was the most radical single transition in all of vertebrate evolution, many terrestrial groups have made the reverse journey, and the slim, sinuous mesosaurs were probably one of the first. Appearing early in the story of reptiles (330–270 million years ago), there is debate about whether mesosaurs had to return to land to lay their eggs as turtles and crocodiles must today, or whether some species gave birth to live young.

**Zdeněk Burian (1905–1981), *Mastodonsaurus*, from *Prehistoric Animals*, 1956.**

Burian was a major twentieth-century "paleoartist" who was able to combine scientific accuracy with an evocative style. In many of his pictures, including this depiction of the large (16ft/5m), 240-million-year-old amphibian *Mastodonsaurus*, the mise-en-scène was as important as the leading actor. Here our chunky protagonist stands stoically in his suitably dank and archaic environment.

**Illustration by Rudolph F. Zallinger (1919–1995) from Jane Werner Watson's *Dinosaurs and Other Prehistoric Reptiles*, 1960; Sleeping *Tyrannosaurus*.**

One might not think of nature's most famous killer having unguarded, almost intimate moments, but presumably *Tyrannosaurus* had to sleep, and a rain storm seems as good a time as any.

# MARY LEAKEY (1913–1996)

## Consuls, Nutcrackers, and Footprints in the Ash

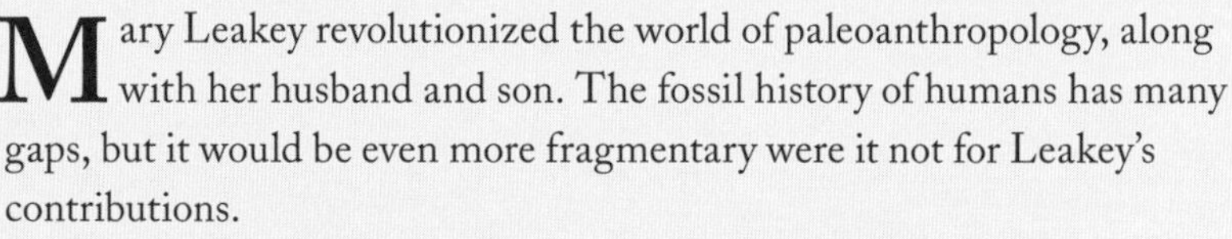

Mary Leakey revolutionized the world of paleoanthropology, along with her husband and son. The fossil history of humans has many gaps, but it would be even more fragmentary were it not for Leakey's contributions.

**Excavation of a hominid footprint trail fossilized in volcanic ash, found by Mary Leakey's expedition to Laetoli, Tanzania, in 1978.**

Born in London, Mary Nicol traveled widely with her family, including to the Dordogne in France where she became fascinated by the evidence of local human habitation dating back 2.5 million years. She married Louis Leakey in 1936 and they moved to Kenya, which was to be their home for the rest of their lives. There they discovered ancient rock paintings, pottery shards, and hand axes at a number of sites. Except when their work was temporarily disrupted by local political upheavals, Mary and Louis continued to excavate, focusing eventually on two sites in nearby Tanzania—the now-famous Olduvai Gorge, where a stream has eroded through layers of sediments deposited within the last few million years and, thirty miles to the south, Laetoli.

Mary had a natural aptitude for finding fossils. In 1948, she discovered a specimen of *Proconsul africanus*, a 25-million-year-old ape which lies near the common origin of great apes and humans (its name means "before Consul"; Consul was a chimpanzee at London Zoo). In 1959 came 1.75-million-year-old *Zinjanthropus boisei*, which the Leakeys nicknamed "Nutcracker Man" due to his massive jaws and teeth. *Zinjanthropus*'s status has changed over the years, and is now often referred to as *Paranthropus boisei* or *Australopithecus boisei*. Emphasizing how modern humans' near-ancestors often coexisted with these

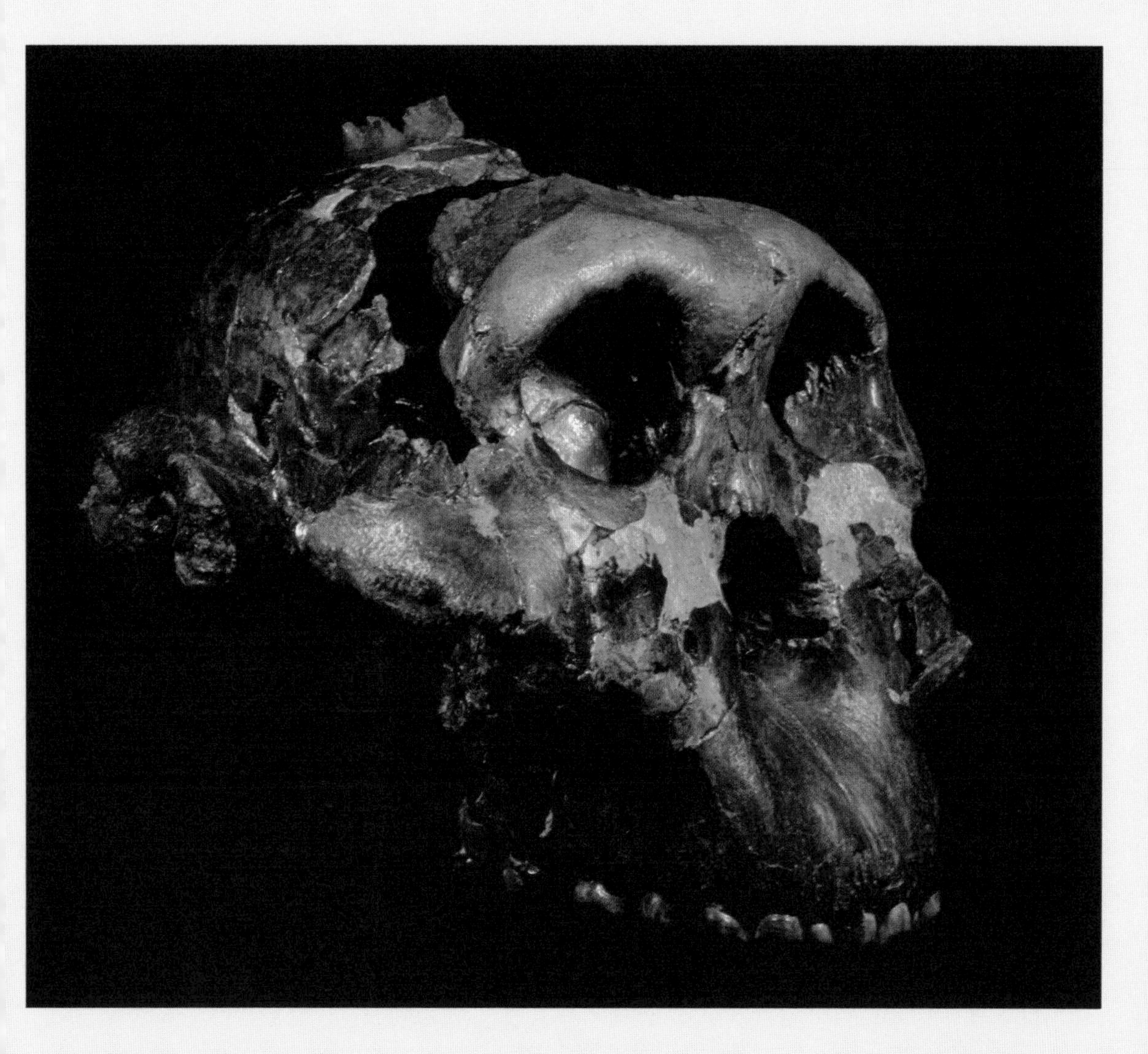

closely related species, in 1960 and 1972 the Leakeys found two specimens of *Homo habilis* (handy man), which were 1.9–2.0 million years old, in the same region.

**Skull of *Zinjanthropus* (now often *Paranthropus* or *Australopithecus*) *boisei*—or "Nutcracker Man"—found during Mary Leakey's expedition to Olduvai Gorge, Tanzania, in 1959.**

In 1978, Mary made probably her most famous discovery: footprint trails of the obviously bipedal *Australopithecus afarensis* (see page 176) in freshly fallen volcanic ash at Laetoli. The prints are thought to date from 3.66 million years ago and show that walking on two legs evolved early in humans' ancestors and their relatives—before large brains, for example—and thus may have been the key innovation that kickstarted our species' remarkable story. Two of the trails appear to be of individuals walking side by side, and a third *Australopithecus* seems to have stepped carefully in their footprints in the ash.

NEAVE PARKER

**Neave Parker (1910–1961),**
***Megalosaurus*, 1960.**

Neave Parker was an English paleoartist who created many images for the publications of the London Natural History Museum, often in collaboration with its resident paleontologists. His dinosaurs have a muscular, slightly glistening quality and are often in beige, muddy brown, or slate gray. His *Megalosaurus* has a strikingly dynamic stance for an image created before the late-1960s "Dinosaur Renaissance," although its neck is shorter and straighter than is now thought to have been the case.

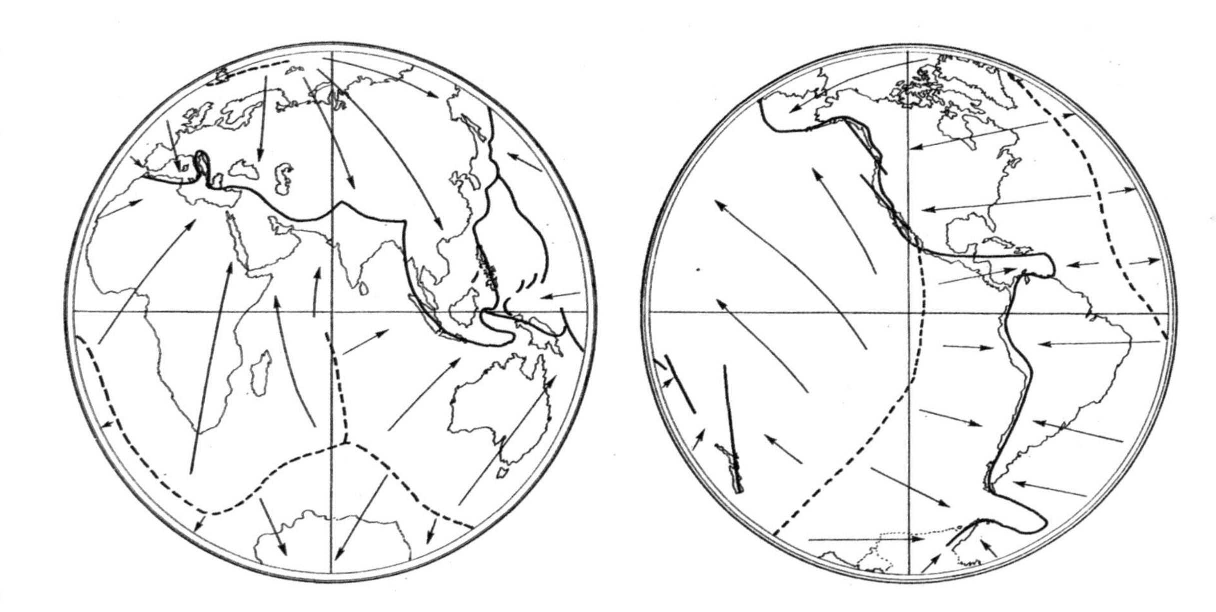

**John Tuzo-Wilson (1908–1993), "Evidence from ocean islands suggesting movement in the Earth,"** ***Philosophical Transactions of the Royal Society of London*****, Series A, vol. 258, 1965.**

Unlike many major advances in science, the theory of plate tectonics—the mechanism underlying continental drift (see page 125)—did not spring from the mind of one thinker, but rather coalesced, almost by committee, in the mid-1960s. Many geologists from around the world, including some who had doubted Alfred Wegener's idea that continents migrate across the Earth's surface, developed a new consensus that this phenomenon is, in fact, real and driven by convection currents in the fluid mantle which underlies the crust. This diagram is from a paper which notes that oceanic islands tend to be older the farther they lie from mid-ocean ridges, and that islands near sea-floor trenches are paradoxically elevated. Together these observations suggest that oceans widen due to the creation of new crust at mid-ocean ridges, whereas crust disappears into the depths of the Earth at ocean trenches—a phenomenon which creates a nearby "bulge" that temporarily raises islands.

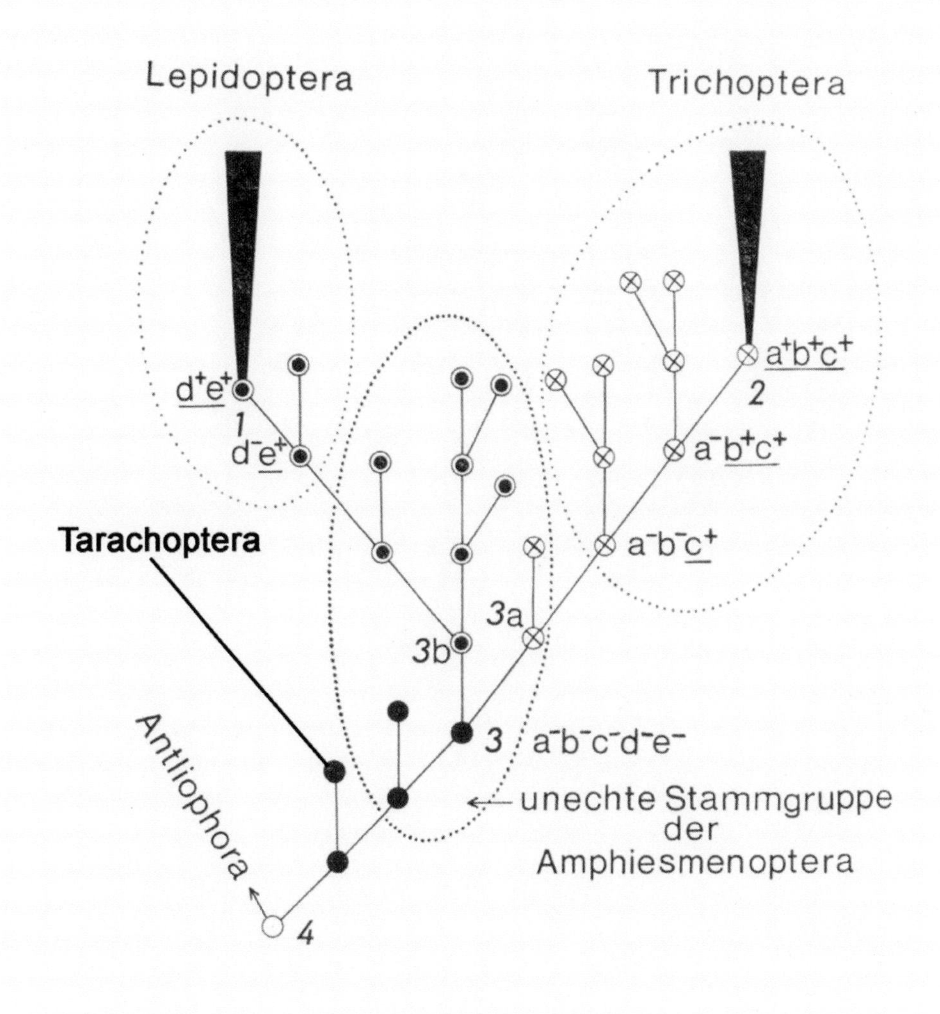


**Willi Hennig (1913–1976), *Die Stammesgeschichte der Insekten*, 1969; A presumed phylogeny of the Amphiesmenoptera (an insect superorder).**

The German biologist Willi Hennig was the originator of the modern system of classification, now usually called "cladistics"—an almost arithmetic scheme for creating evolutionary "trees" according to strict rules relating to the appearance of novel characteristics. He started to develop his ideas as a prisoner of war after the First World War, and published them in the 1950s and 1960s. Suddenly, the assumptions and subjectivity of previous systems of classification disappeared, to be replaced by evidence-based scientific hypotheses presented in austere stick-tree form.

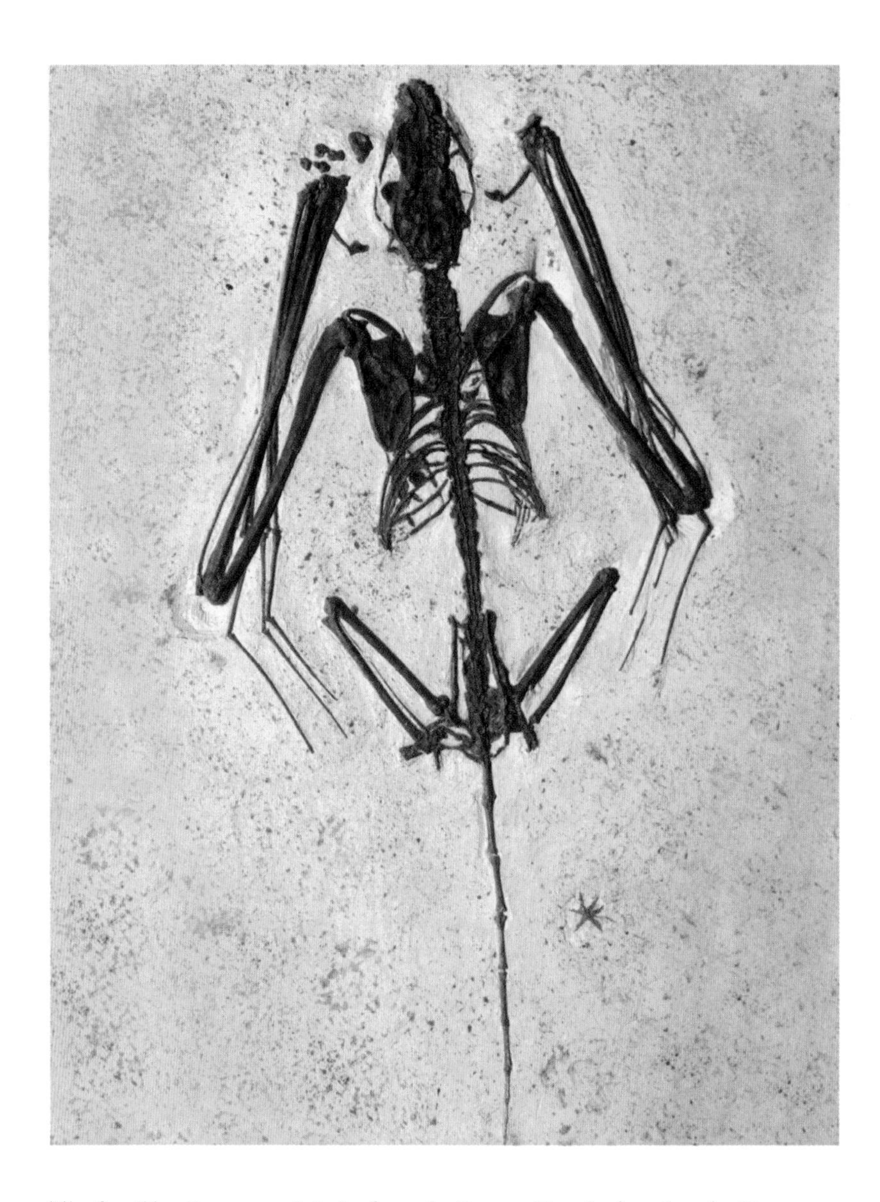

**The fossil bat *Icaronycteris index* from the Eocene Epoch, found at the Green River Formation, Wyoming.**

In a strange parallel with frogs (see page 149), the distinctive body plan of bats appears suddenly in the fossil record in strikingly "final" form, apart from the retention of a long tail. Thus we assume that bats either evolved remarkably quickly around 50 million years ago, or transitional forms did not fossilize in locations searched by paleontologists. Representing only the third time powered flight has evolved in vertebrates, bats have become the second most species-diverse mammalian group, behind rodents.

**Rudolph F. Zallinger (1919–1995), Detail from *The Age of Mammals* fresco, Peabody Museum of Natural History, 1961–1967.**

A Yale graduate, Zallinger taught in that University's school of fine arts, later becoming the Peabody Museum's artist-in-residence. He created many depictions of ancient life for institutions and popular publications, but the most famous are the murals at the Peabody. This image is a detail from the huge *Age of Mammals* fresco—measuring 6 x 60ft (1.8 x 18m)—from the south wall of the Hall of Mammalian Evolution. Zallinger took some artistic liberties with the number of animals in his pictures, so they often seem to be jostling for space, but he was otherwise careful to be faithful to contemporary scientific ideas about the animals, plants, landscapes, and climates he depicted.

**Top: Matteo De Stefano, Reconstruction of *Dunkleosteus terrelli*, Museo delle Scienze di Trento, Italy. Above: The late-Devonian fossil placoderm *Dunkleosteus*, Royal Tyrrell Museum, Drumheller, Alberta, Canada.**

Marauding Devonian seas from 380–360 million years ago, the genus *Dunkleosteus* contained the largest known placoderm (plate-skin) fish—a numerous and diverse group closely related to modern jawed vertebrates. Some species grew to over 25ft (7.6m) and probably weighed more than 3 tons. Some fossils show females containing developing embryos—the first known example of internal gestation of offspring in backboned animals.

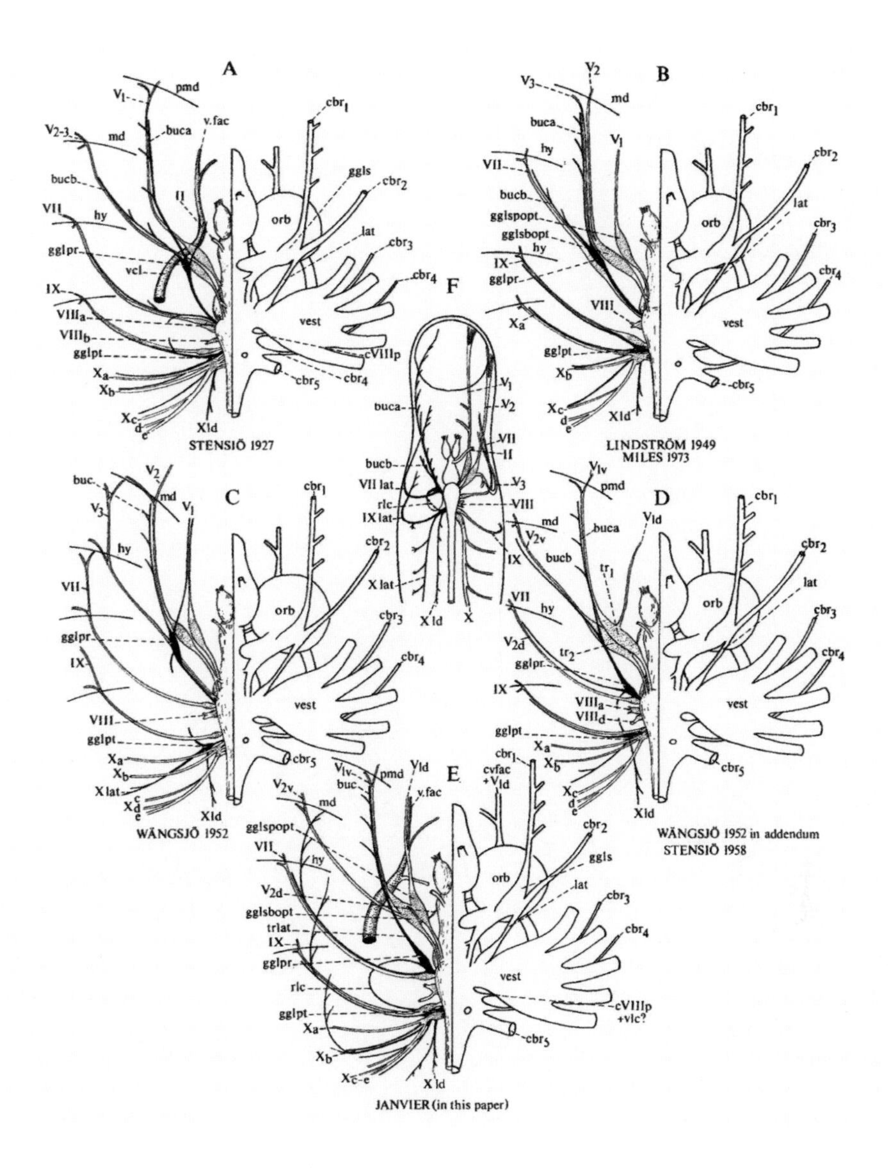


**Philippe Janvier, "The Sensory Line System and Its Innervation in the Osteostraci (Agnatha, Cephalaspidomorphi)," *Zoologica Scripta*, vol. 3, 1974; Ventral view of the brain, the main cranial nerves, and the lateralis nerve complex in osteostracans.**

Many ancient fish had such robustly armored skull-carapaces that they fossilized well—so well, in fact, that they can be sawn into slices, so allowing the course of delicate internal canals to be traced. This yields a three-dimensional model of the brain and its associated nerves, and because those nerves follow a strikingly similar scheme in different vertebrate groups, they can be used to trace evolutionary affinities and morphological trends between groups whose relationships would otherwise be lost in the mists of time. The "osteostracans" pictured here were jawless fish, probably related to modern lampreys (labeled F, center), and it is tempting to speculate that their bony armor evolved in response to the ferocious-looking placoderms pictured opposite.

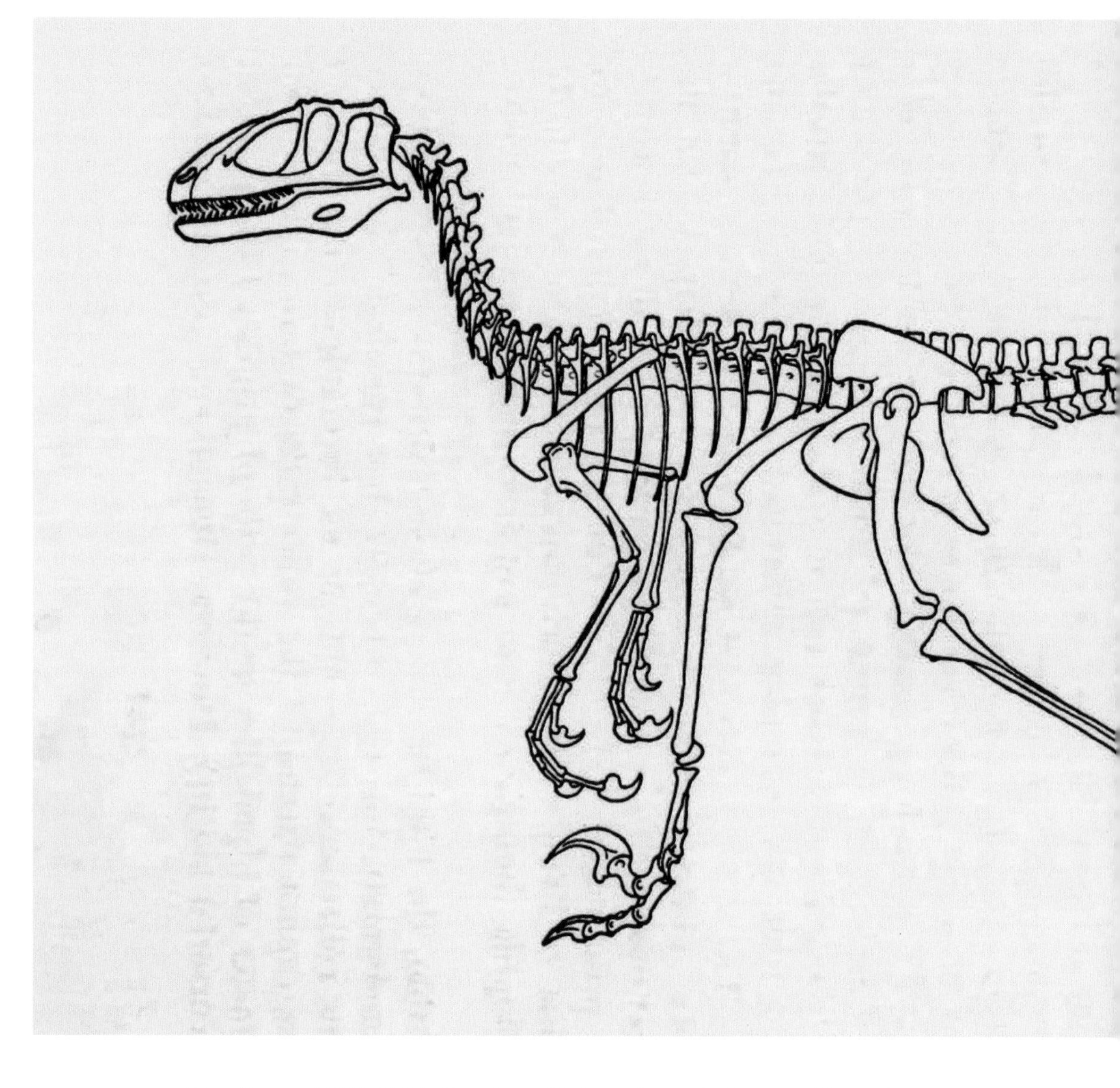

**John Ostrom (1928–2005), *Osteology of Deinonychus antirrhopus, an Unusual Theropod from the Lower Cretaceous of Montana*, 1969.**

Related to the smaller but now more famous *Velociraptor* (see page 136), *Deinonychus* was the bipedal theropod (beast-foot) dinosaur which contributed most to the late-1960s' "Dinosaur Renaissance" (see page 132). The paleontologist John Ostrom argued that the skeleton of *Deinonychus* was obviously that of a nimble, speedy hunter completely at odds with the prevailing image of dinosaurs as cold-blooded sluggish behemoths—and Robert Bakker's powerful diagrams seemed incontrovertible visual proof of this paradigm shift. Over the following years our understanding of dinosaur biology changed completely; *Deinonychus* is, for example, now thought to have been feathered.

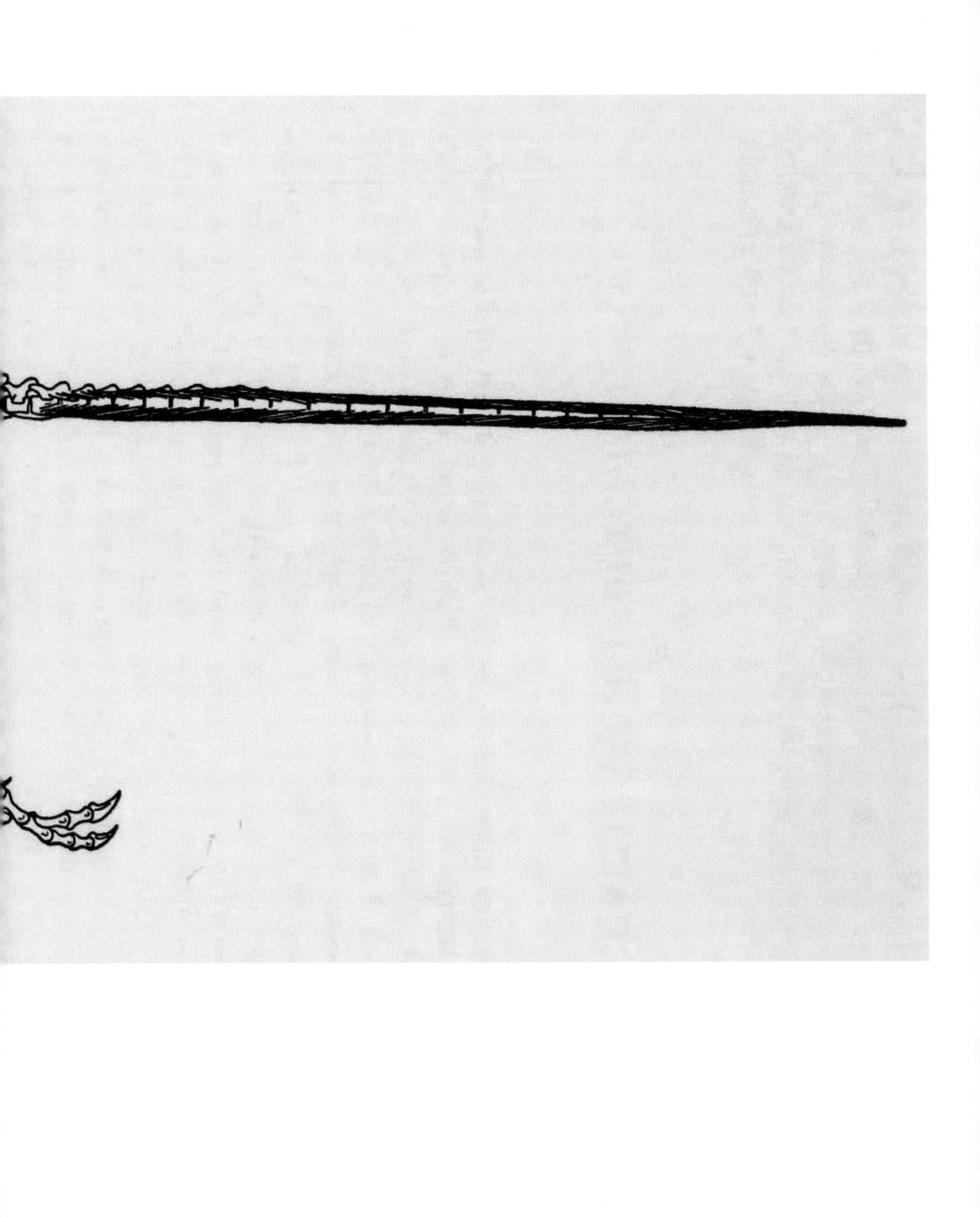

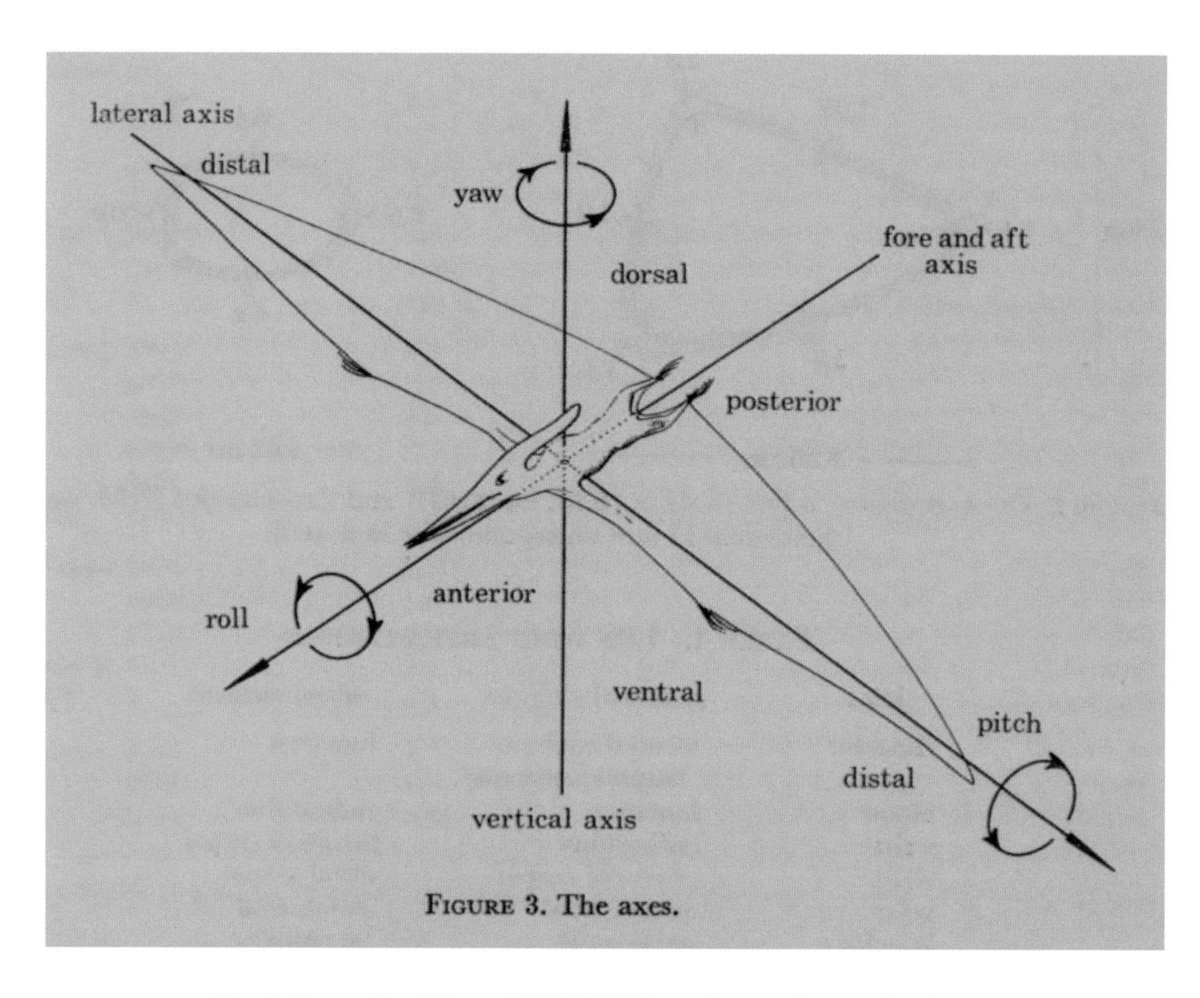


**Cherrie Bramwell and George Whitfield, "Biomechanics of Pteranodon," *Philosophical Transactions of the Royal Society of London*, Series B, vol. 267, 1974.**

Few ancient creatures have inspired scientists as much as the fragile pterosaurs. The image above is from a study of the mechanics of *Pteranodon ingens* (immense wing-tooth) with a wingspan of 23ft (7m) but an estimated weight of only 37lb (17kg). The authors proposed that *Pteranodon* was a "superb low speed soaring aircraft" which could fly at 18 miles per hour, and thus become airborne from a standing start in a stiff breeze or by launching gently from a cliff. The apparent fragility of *Pteranodon* must surely have been compounded in larger, more recently discovered forms such as *Quetzalcoatlus* (shown opposite), for which wingspan estimates range from 35ft (11m) to a less realistic 70ft (21m).

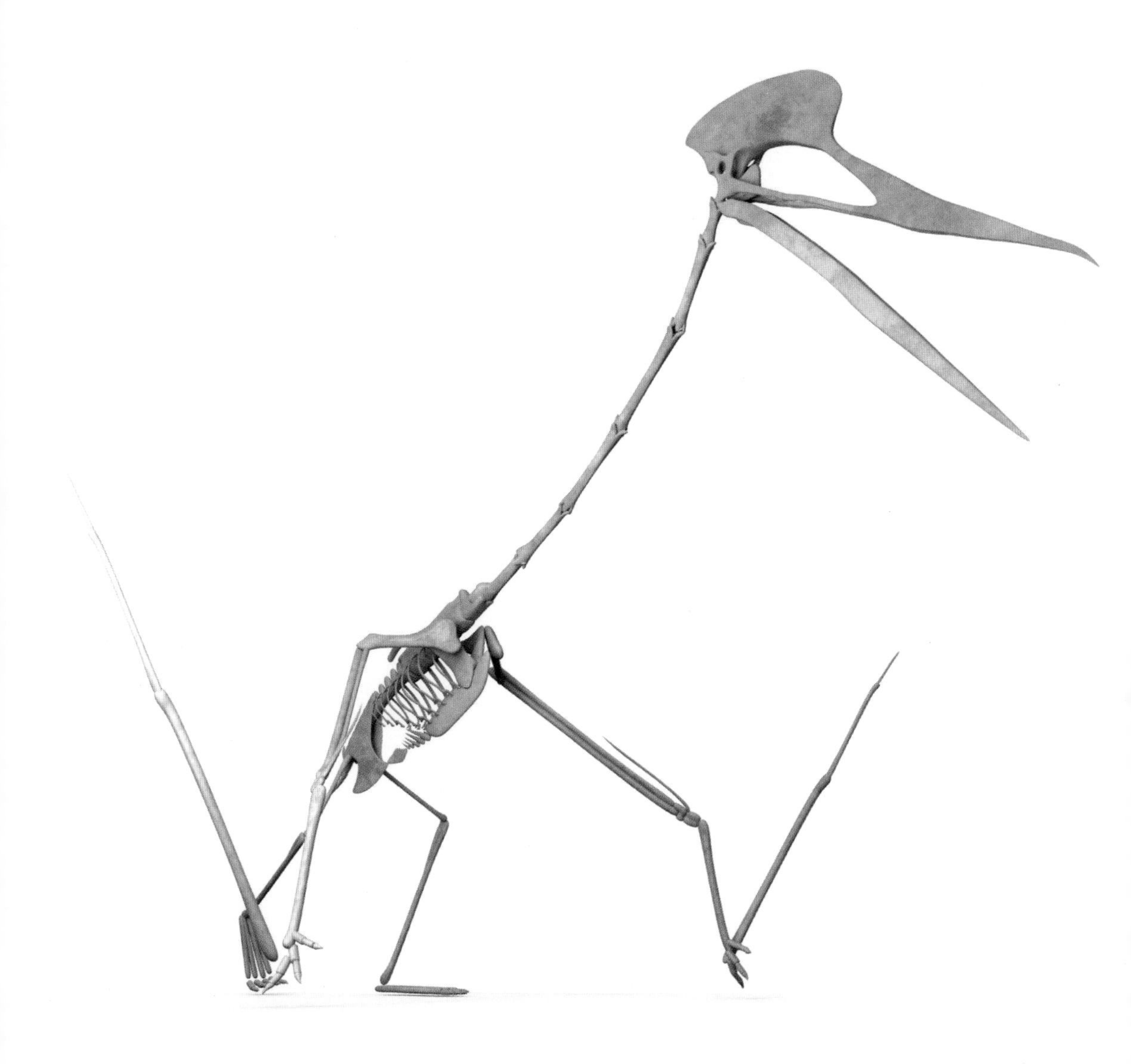

**Roger Harris, Reconstruction of a *Quetzalcoatlus* skeleton.**

Discovered in Texas, the genus *Quetzalcoatlus* probably contained the largest animals ever to have flown. Its size and posture were initially inferred from fragmentary fossil evidence, but some species may have stood as tall as a giraffe when on the ground, and in flight their wingspan probably exceeded 35ft (11m)—and some earlier estimates reached 70ft (21m).

**Partial skeleton of *Australopithecus afarensis*—known as "Lucy"—discovered at Hadar, Ethiopia, in 1974, National Museum of Ethiopia, Addis Ababa.**

**Élisabeth Daynès, Reconstruction of *Australopithecus sediba*, University of Michigan Museum of Natural History.**

Discovered in 1974 in the Afar Depression in Ethiopia, the *Australopithecus afarensis* specimen known as "Lucy" (opposite) is the most famous of all hominin fossils. Named after a Beatles song played repeatedly in the expedition camp, Lucy dates from approximately 3.2 million years ago and possesses a mixture of human and great ape traits—she seems to have walked bipedally and upright, but her brain was of a similar size to a modern chimpanzee, while her pelvis appears adapted for a birth process intermediate between those of modern humans and chimps. Paleopathological investigations have identified fresh fractures suggesting Lucy may have died after falling from a tree. Several *Australopithecus* species have now been identified, including *A. sediba* (above) from South Africa. Although their interrelationships are tangled, one of them may be the ancestor of humans, while later *Australopithecus* species coexisted with members of the nascent *Homo* lineage.

# LUIS AND WALTER ALVAREZ
## (1911–1988 AND 1940–)

## Did an Extraterrestrial Impact Kill the Dinosaurs?

At the end of the Cretaceous Period 66 million years ago, something caused the mass extinction of approximately three-quarters of all animal species, famously including all dinosaurs except birds and paving the way for mammals to become our planet's dominant animals.

For decades no one knew what caused this "K-T extinction event," until evidence started to accumulate around 1980 that it may have had a celestial explanation. One of the first hints was the discovery of globally increased concentrations of the element iridium, usually rare on Earth, in a thin band of sediments dating from the "K-T boundary." Iridium is found at far higher concentrations in asteroids than in the Earth's crust, and some sediments show more than a hundred-fold increase. This led the Nobel-Prize-winning physicist Luis Alvarez and his son Walter to propose that it was an asteroid impact which caused the K-T extinction.

**Diagram by Jake Bailey and David Kring for NASA/University of Arizona Space Imagery Center, showing cenotes (sinkholes) near Chicxulub, Yucatán Peninsula, Mexico, 2000.**

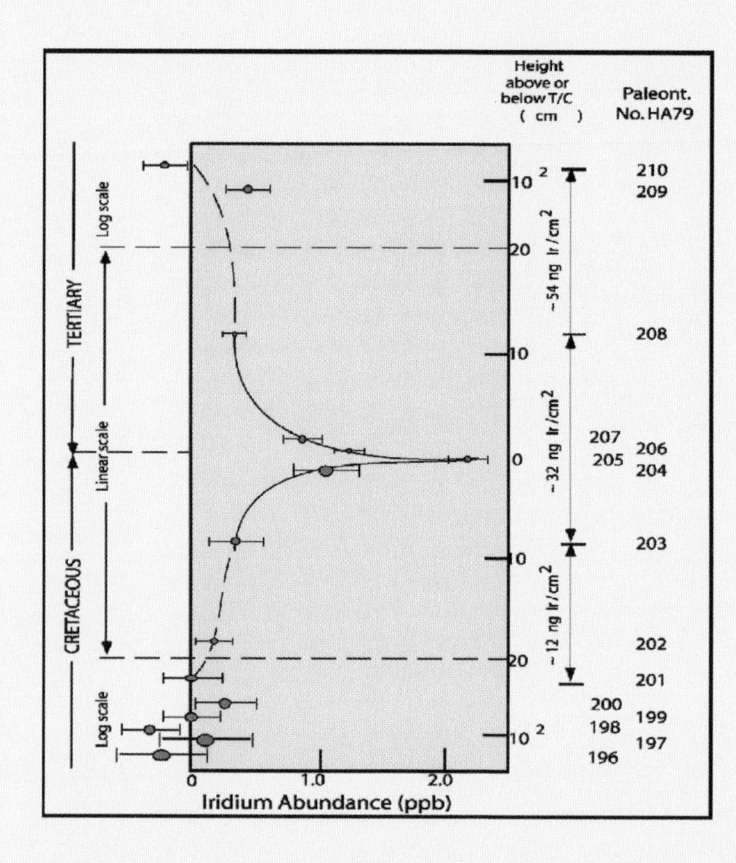


**Florentin Maurrasse and others, "Spatial and temporal variations of the Haitian K/T Boundary record,"** ***Journal of Iberian Geology,*** **vol. 31, 2004; Iridium anomaly at the K/T boundary layer stratotype section.**

Gradually more evidence began to emerge, including the discovery of a 110-mile-wide impact crater intersecting the coastline near Chicxulub in Mexico. There, geologists have now discovered melted and "shocked" minerals, an anomaly in the strength of gravity, evidence of seabed scouring and tsunami backwash, as well as space-based radar evidence of a huge, buried, bowl-shaped structure. Also, a curved buried ridge has blocked the flow of groundwater, creating a tell-tale arc of cenotes, or sinkholes (see map opposite).

So asteroid impact is now the favored theory for the cause of the K-T extinction. It is thought an asteroid of ten to thirty miles diameter impacted obliquely at 30,000 miles per hour, causing a plasma plume, supersonic winds, a seismic wave equivalent to a magnitude-10 earthquake, tsunamis hundreds of feet high, a spray of molten ejecta, a worldwide covering of dust, and a long-term global "impact winter." And remarkably, the K-T extinction is just one of several such events now known to have occurred—indeed, an even more profound extinction event occurred 252 million years ago, often called the "Great Dying" (see page 192).

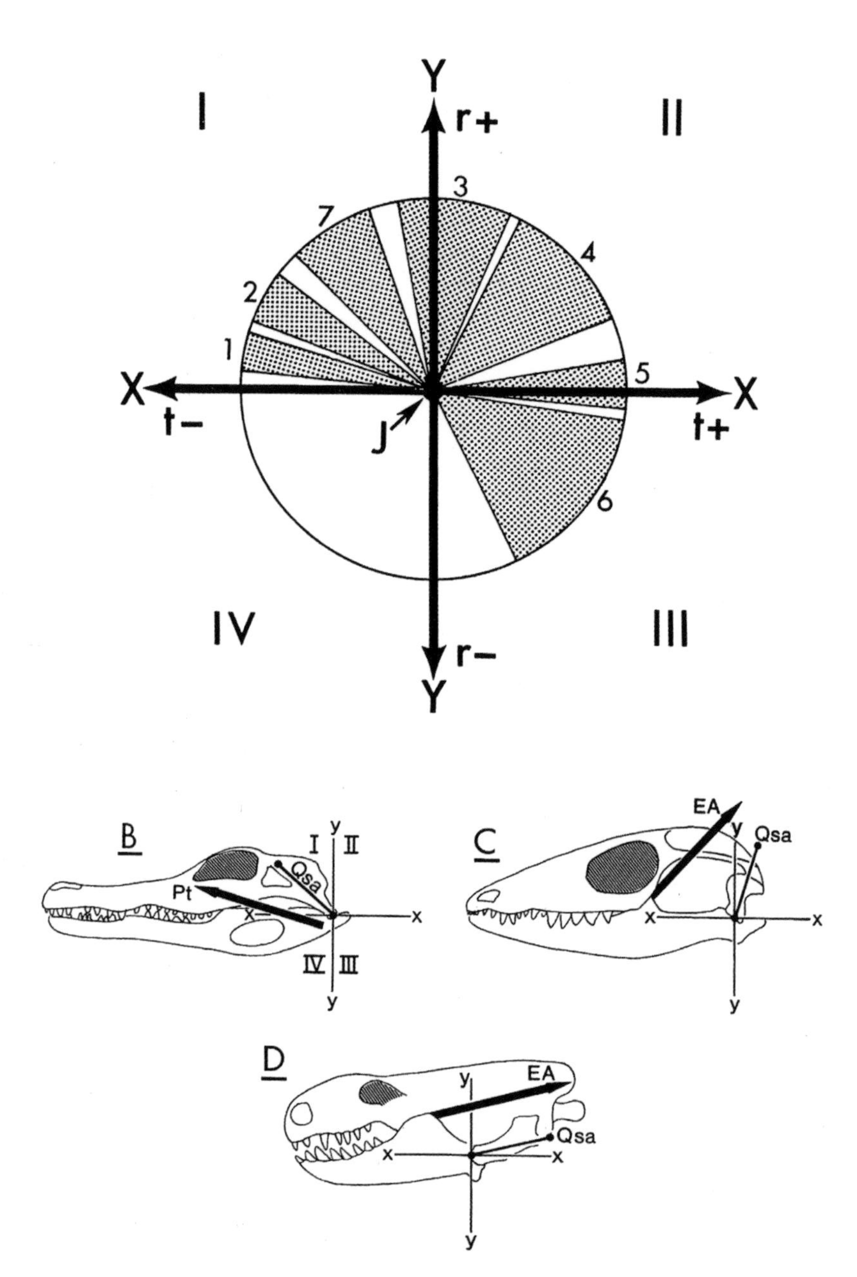


**Dennis Bramble, "Origin of the mammalian feeding complex: models and mechanisms,"** ***Paleobiology*****, vol. 4, 1978; Diagrammatic figures of the skulls and jaws of (B) Crocodilian, (C) Lizard, and (D) Amphisbaenian.**

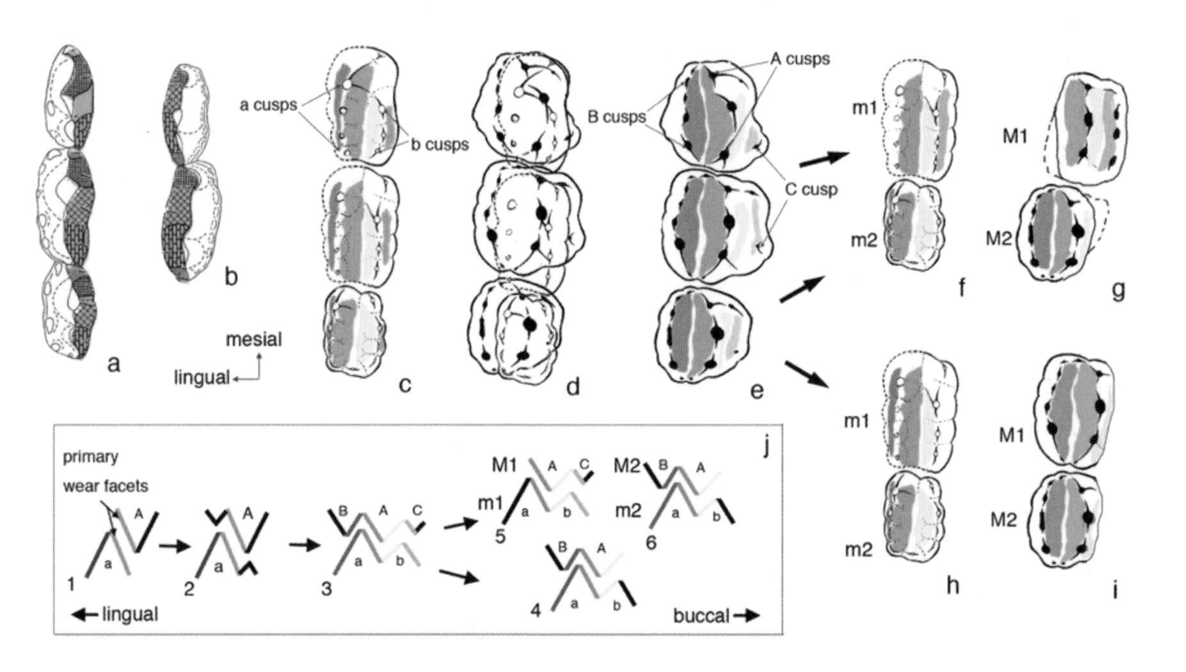


**Jin Meng, "Mesozoic mammals of China: Implications for phylogeny and early evolution of mammals," *National Science Review*, vol. 1, 2014. A hypothesis for transformation of the allotherian tooth pattern from a "triconodont" tooth pattern (schematic views of the molar tooth crowns of early mammals).**

It could be argued that the success of mammals is based largely on their ability to chew—to divide food into tiny fragments which can be speedily and efficiently digested to provide the energy and nutrients required for these creatures' high-energy lifestyles. This adaptation comprises two main elements: jaws and teeth. The diagram on the page opposite is from an influential analysis of jaw evolution in the lineage leading to modern mammals. It demonstrates how proto-mammals' jaw muscles became increasingly complex to allow them to exert surprisingly large forces on food items without damaging the mandible or the jaw joint. Mammals also possess neatly aligned ranks of exquisitely sculpted teeth which intermesh to slice food efficiently—and teeth usually fossilize better than any other tissue, so their progressive modification has been an excellent way to trace the evolution of the ancestors of modern mammals.

Chapter Four

# THE MODERN ANCIENT

1980–PRESENT

Giant "X-fish" *Xiphactinus molossus*, measuring 16ft (5m), from the Cretaceous Period and found in the Western Interior Seaway of North America in 1982.

# THE MODERN ANCIENT

## 1980–PRESENT

In the last forty years paleontology has become both broader, merging with and drawing from cognate sciences, as well as larger—there are now more active paleontologists in the world than ever before. The sheer volume of research has increased as has its quality, so it is fair to claim that we are now in the science's greatest golden age. As a result, this chapter can contain only a small selection of the insights and innovations made by a veritable army of recent paleontologists, now allied with the fields of paleopathology, forensic reconstruction, evolutionary theory, diagnostic imaging, aerodynamics, hydrodynamics, environmental science, climatology, and even satellite imaging.

Yet in the 1980s there was a very real challenge to the classical model of paleontologists plucking fossil animals from the ground and poring over their anatomy. Rapid advances in molecular biology meant that for the first time DNA codes could be readily sequenced. DNA is the long linear molecule present in most body cells which contains the digital codes organisms use to develop, survive, and reproduce—humans, for example, have roughly 20,000 such codes. And because closely related species have more similar DNA than distantly related ones do, we can now compare

**Skull of the Late Devonian Era tetrapod *Acanthostega*.**

species' DNA codes to determine their interrelationships. This "molecular taxonomy" has burgeoned since, but it certainly has not replaced "bones-and-teeth" paleontology as many had feared. First of all, DNA can only be obtained from a few fossils, because it gradually degrades over time. Second, convergent evolution—the process by which animals share features because they face similar challenges rather than because they are truly related—is as much a confusing factor when comparing DNA as when comparing bones. And third, DNA does not (yet) give us much information about animals' shapes, environments, or ways of life.

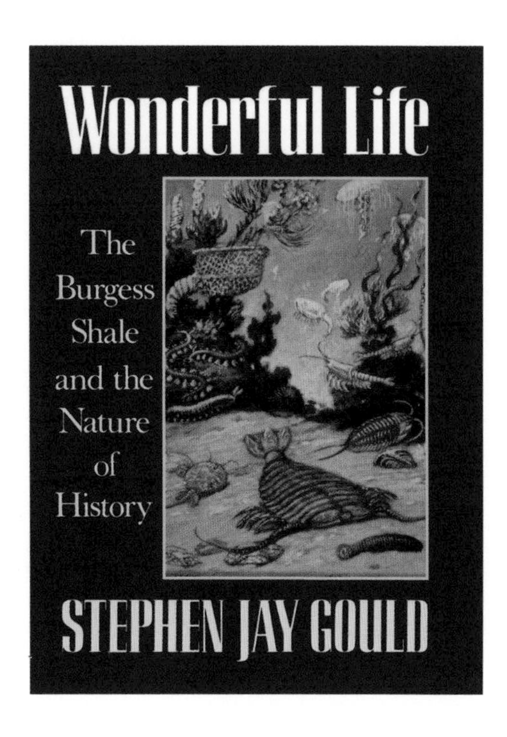


**Stephen Jay Gould, *Wonderful Life, The Burgess Shale and the Nature of History*, 1989.**

So traditional paleontology continues, and indeed has been bolstered by fossil discoveries from new and under-exploited regions of the world. Fossilization is a rare, chance event, so limiting excavation to just a few locales would restrict our knowledge to only an unrepresentative subset of the environments and animal types among the biological profusion of the past. Remarkably preserved finds are now regularly made in China and North Africa, dinosaurs have been excavated in Antarctica, and paleontologists trek through the wilderness of Greenland and Arctic Canada.

In recent decades these latter, northern Lagerstätten have added especially to our understanding of the fish-tetrapod transition, the process by which vertebrates evolved from fully aquatic ancestors to four-footed (tetra-pod) land-living forms. Two particular stolid-looking beasts, *Acanthostega* (see page 194) and *Tiktaalik* (see page 212), have joined *Ichthyostega* (see page 147) in the pantheon of our ancient, fresh-legged forebears. Whereas the move to land seems relatively unproblematic for crustaceans and their kin, it was the most structurally profound change ever to occur in backboned animals, necessitating the reconfiguration of fins into jointed limbs, and the complex repurposing of gills, noses, and swim bladders for life above the surface. Vertebrate morphologists still debate the evolutionary origins of wrists, eardrums, and nostrils, while embryologists argue about how the developmental program of fishes was modified to create something which could clamber about and raise its head above the undergrowth. In fact, many now claim the changes we think of as assisting our ancestors' move to dry land actually evolved to aid underwater activities, such as striding purposefully across the bottom of

**The Pleistocene Epoch frozen baby mammoth "Lyuba", found in the Yamal Peninsula, northern Siberia, in 2007.**

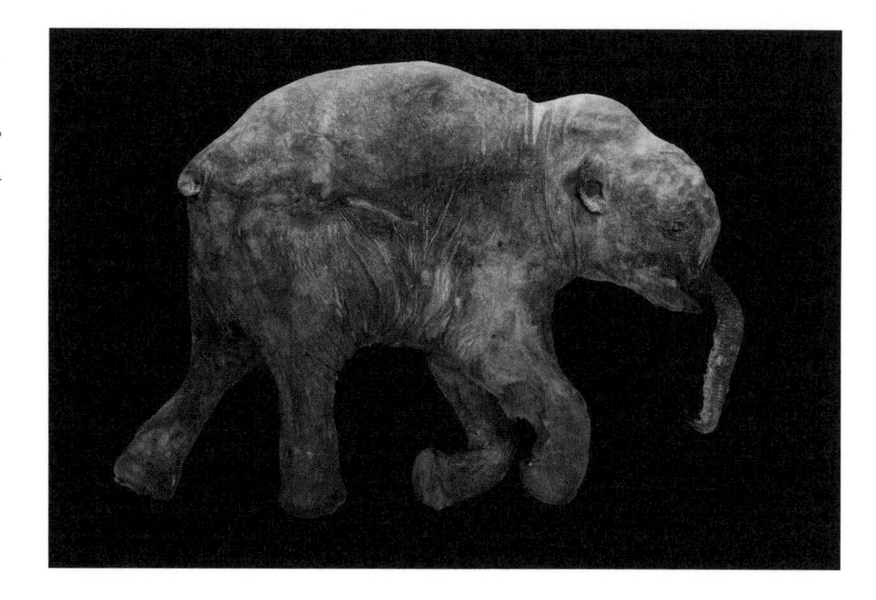

ancient bodies of water. Indeed, the fish-tetrapod transition seems to have been a very leisurely process, and takes up a large fraction of tetrapods' time on Earth—starting at least 400 million years ago.

There have also been increasingly animated scientific discussions about the evolutionary process itself. Disregarding the futile distractions of creationism and its pseudoscientific rearguard battalion, "intelligent design," much remains to be discovered about how natural selection causes evolution to take place. The 1989 book *Wonderful Life* by evolutionary biologist Stephen Jay Gould (1940–2002) not only gave the strange creatures of the Burgess Shales the fame they had always deserved (see page 118), but also re-catalyzed arguments about whether evolution necessarily drives life toward ever-increasing complexity. In addition, active research is also underway into how quickly or slowly evolution occurs, and whether its rate is constant or variable, as well as how a single species actually splits into two.

Around the same time, our view of the history of life took on a wider perspective as the search for life spread out into the Solar System. In 1996, the announcement was made of microbial fossils embedded in the Allan Hills 84001 meteorite, a chunk of rock blasted from the surface of Mars which subsequently landed in Antarctica. Although the interpretation of these "fossils" as relics of Martian organisms was subsequently discounted, the find did demonstrate how meteorites might convey organisms from one celestial body to another, possibly even suggesting an extraterrestrial origin for life on Earth. In addition, it now seems likely that several of the Solar System's planetary satellites and dwarf planets possess large subsurface oceans in contact with underlying rocks heated by each sphere's inner processes—a combination of factors already known to be conducive to life in Earth's ocean depths.

One might assume that the worldwide, science-focused nature of modern paleontology would lead to a growing consensus about the history of life on Earth, so that few major surprises await us. However, paleontologists are entirely dependent on a raw material that must be physically extracted from the ground, and fossilization is a rare, capricious, and sometimes localized event. A good example of a perspective-changing discovery is the unexpected 2003 unearthing of *Homo floresiensis*, "Flores Man" or the "hobbits" of popular media, a diminutive close relative of modern humans (see page 211). We now believe that today's *Homo sapiens* is unusual in being the only human species in existence, whereas for most of the last few million years there has been more than one *Homo* species on the planet. Astoundingly, it is now thought that *Homo floresiensis* inhabited the Indonesian archipelago until only 50,000 years ago or, in the terms used throughout this book, 0.05 million years ago—the blink of a paleontological eye. More recently, fossil and molecular evidence have indicated that another human species or subspecies, the Denisovans, may have inhabited east and southeast Asia even more recently, and may not have become extinct, but instead genetically merged into *Homo sapiens* by interbreeding.

**Vincent Santucci and others, "Crossing Boundaries in Park Management," 2001; Western Interior Seaway of North America during the Cretaceous Period.**

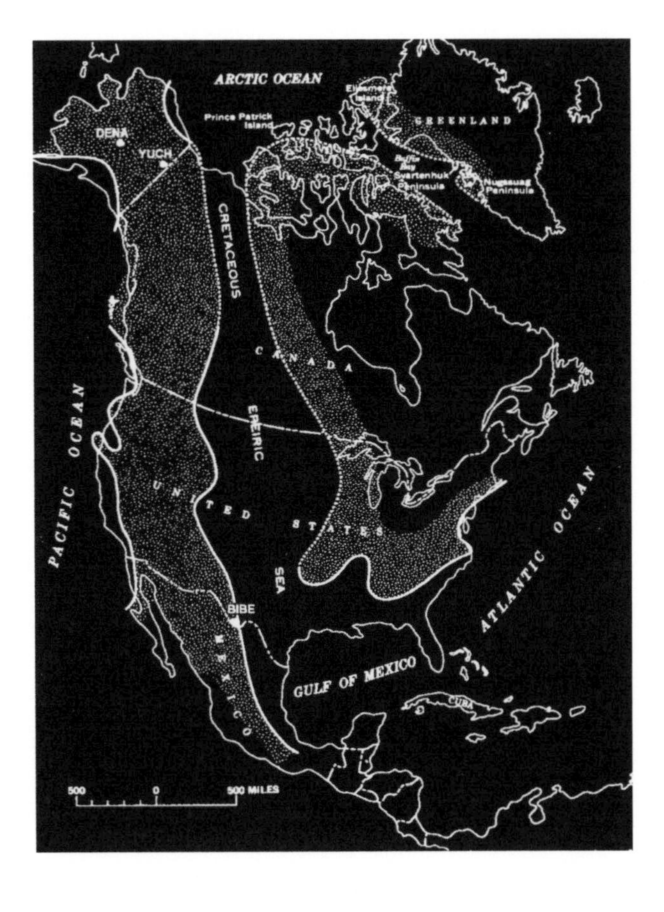


Although paleontology is a hard science, and the data it yields are being held to ever more stringent evidential standards across a widening array of disciplines, its social and historical dimension has also grown in recent years. In some ways the bridge between prehistory and history is being discovered—we now know, for example, that many animals and peoples we only know from

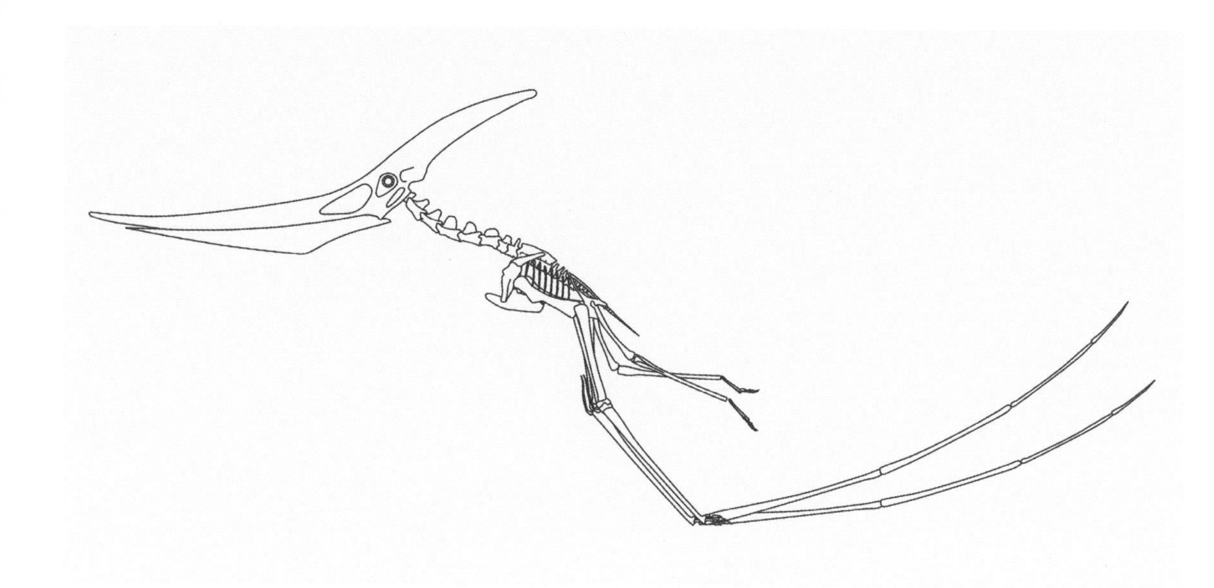

**Mark Witton and Michael Habib, "On the size and flight diversity of giant pterosaurs, the use of birds as pterosaur analogues and comments on pterosaur flightlessness," PLOS ONE vol. 5: 2010; Skeletal reconstruction of a quadrupedally launching *Pteranodon*.**

fossils actually coexisted with humans with brains as large as ours. Moas, mammoths, and hobbits: eyes like ours have seen them all. And it is also increasingly clear that paleontology has actually existed throughout human history, and has blended into folklore and legend, even surviving in fossil-related place names. There are many ancient writings which are thought to describe, rationalize, and appropriate contemporary fossil discoveries, and there is also no reason to believe fossil-finding did not occur before the advent of writing. It has also been suggested that the *Mahabharata*, the world's most epic epic poem, distilled beliefs and myths about giant fossils into its titanic narrative.

And today, the social dimension of paleontology is also undergoing a reappraisal. More than most sciences it relies on discoveries across the globe, once made in the old European empires or more recently in states subject to the post-war Western economic hegemony. For much of the last two centuries fossils were excavated from Lagerstätten, usually by the labor of local people, but spirited away to distant museums which made little reference to the cultural context in which they were found. This is slowly changing, however, as has the fact that the paleontologists of the past were usually white men—fossils are now found and analyzed by diverse international teams, and at least some remain close to where they are discovered.

So fossils live on in legend, culture, and, of course, science. We are now in the greatest golden age of paleontology, a dramatic expansion and acceleration when discoveries and the insights they bring are coming thick and fast. And all of this not a moment too soon. A century ago the stories of extinction and environmental catastrophe recounted by fossils were interesting historical diversions, while now they stand as warnings. After all, those who cannot remember the past are condemned to repeat it.

**"There is grandeur in this view of life, with its several powers, having been originally breathed into a few forms or into one; and that, whilst this planet has gone cycling on according to the fixed law of gravity, from so simple a beginning endless forms most beautiful and most wonderful have been, and are being, evolved."**

Charles Darwin

**The Permian Period fossil amphibian *Sclerocephalus haeuseri*, discovered at Rheinland-Pflaz, Germany.**

**The Cretaceous Period fish *Xiphactinus audax*, discovered in the Western Interior Seaway of North America in 1982.**

For most of the Cretaceous Period, and into the subsequent Cenozoic Era, much of North America was covered by the Western Interior Seaway—a wide inland sea stretching from what is now the Arctic North of Canada to the Gulf of Mexico. As a result, Great Plains fossils from this time reflect a diverse shallow-sea ecosystem, including this specimen from a group of bony fish that could grow up to 16ft (5m) long.

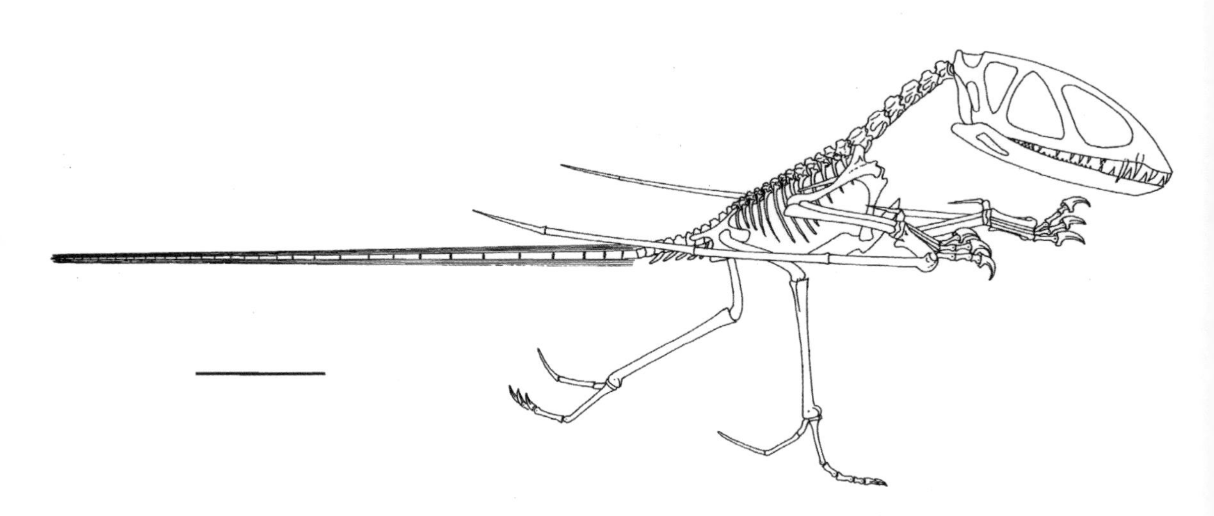

**Kevin Padian, "Osteology and functional morphology of *Dimorphodon macronyx* (Buckland) [Pterosauria: Rhamphorhynchoidea] based on new material in the Yale Peabody Museum," *Postilla*, vol. 189, 1983; Reconstruction of *Dimorphodon macronyx* in bipedal terrestrial progression.**

Flight can evolve in vertebrates in two ways: by running animals acquiring adaptations to increase the height of "upward" leaps, or by climbing animals gaining the ability to glide "downward." In this reconstruction of a small early pterosaur (with a wingspan of approximately 4ft/1.2m) found in 200-million-year-old rocks in England, Kevin Padian makes a clear argument for early pterosaurs as bipedal runners, whose wings had already lost the ability to aid locomotion across the ground.

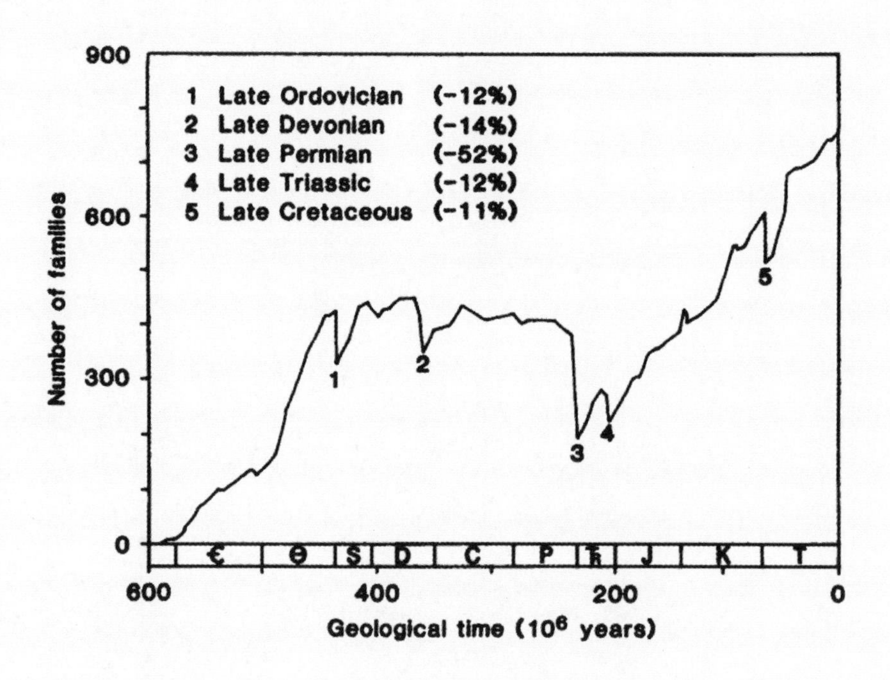


**David Raup and John Sepkoski, "Mass extinctions in the marine fossil record," *Science*, vol. 215, 1982; Standing diversity through time for families of marine vertebrates and invertebrates.**

This graph of marine fossil diversity over time, starting 600 million years ago, shows four or five sudden major declines (the statistical evidence for the second, late-Devonian event was unclear at the time). The final extinction was the fifth, "K-T" event which extinguished the non-avian dinosaurs (see page 178), although this is dwarfed by the severity of the third, late-Permian "Great Dying."

***Ginkgo huttoni* in sandstone, from Scalby Ness, Scarborough, England, Middle Jurassic era. Paläontologisches Museum, Munich.**

Ginkgos are unusual plants dating back 270 million years, and just one species remains today, in China. Thus even this fossil, which grew in Antarctica before that continent moved into its current polar location, lies in the more recent half of this precarious group's history.

# JENNY CLACK (1947–2020)

## Gaining Ground

The English paleontologist Jenny Clack was an expert on the fish-to-tetrapod transition—vertebrates' move from water to land between 400 and 350 million years ago, a transition that she argued was gradual and complex.

Fascinated by fossils throughout her life, Clack wanted to fill in the gaps in the story of the move onto land—a frustrating interruption in the fossil record often called "Romer's Gap." She had realized there were specimens of early tetrapods in museums which had not been identified, and also that some earlier reconstructions required revision. Her chance discovery in the Cambridge Department of Earth Sciences of some tetrapod fossils from Greenland inspired her to lead the first of several

**Jenny Clack (1947–2020), the late-Devonian tetrapod *Acanthostega*.**

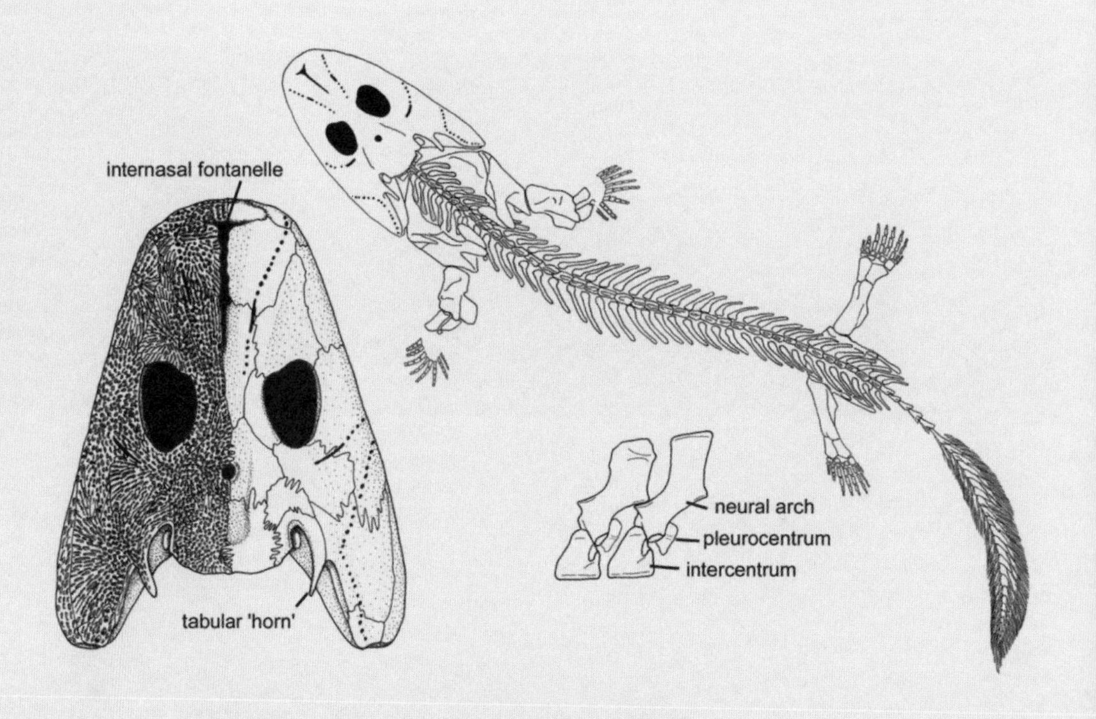

fossil-hunting expeditions there in 1987, along with her husband Rob, whom she had met at a Birmingham motorcycle club a decade earlier.

The Greenland expeditions uncovered an Aladdin's cave of early tetrapods, including new fossils of *Ichthyostega* (see page 147) and excellent specimens of the apparently more-aquatic *Acanthostega*. Rather than a simple narrative of a single species hauling itself decisively onto dry land, what emerged was a story lasting tens of millions of years in which a number of species adapted to a variety of ecological niches—walking along riverbeds, snapping at prey at the banks of watercourses, and some, like *Crassigyrinus* (see page 196), reverting to a fully aquatic mode of life. Many species combined fish-like gills and tails with tetrapod-like limbs and spines, although it now seems that many of the features we think of as adaptations for life on land may have evolved, sometimes more than once, in creatures with an almost entirely aquatic habit.

Jenny Clack was to become professor of vertebrate palaeontology at Cambridge and was immensely influential. She published *Gaining Ground, The Origin and Evolution of Tetrapods* in 2002, and taught both Paul Upchurch and the author.

**Professor Jenny Clack in the University of Cambridge Museum of Zoology, 2012. Photograph Maja Daniels.**

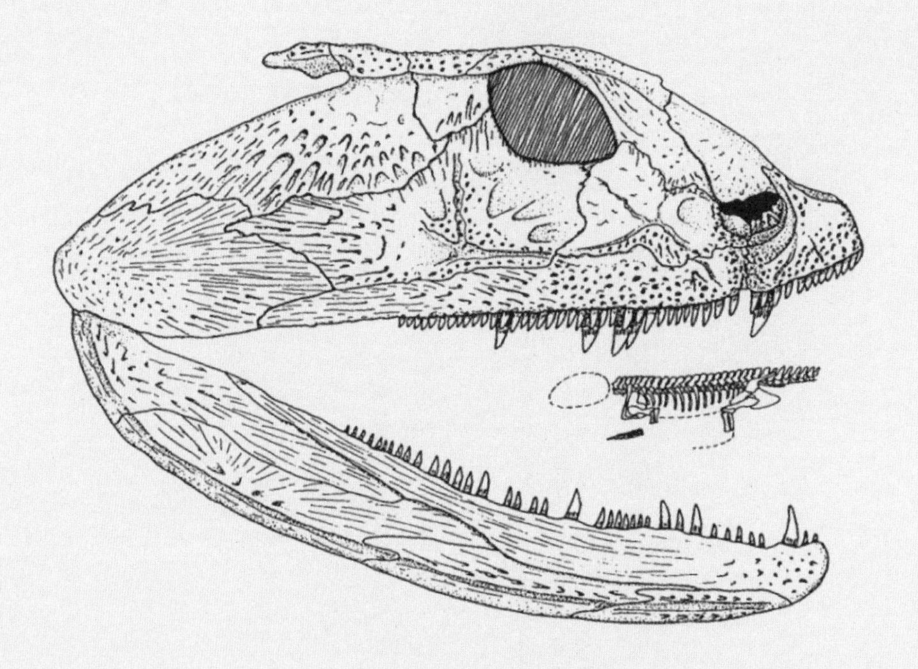

**Above: Jenny Clack (1947–2020), *Crassigyrinus scoticus* (with *Casineria kiddii* in its mouth), found in Scotland in 1987.**

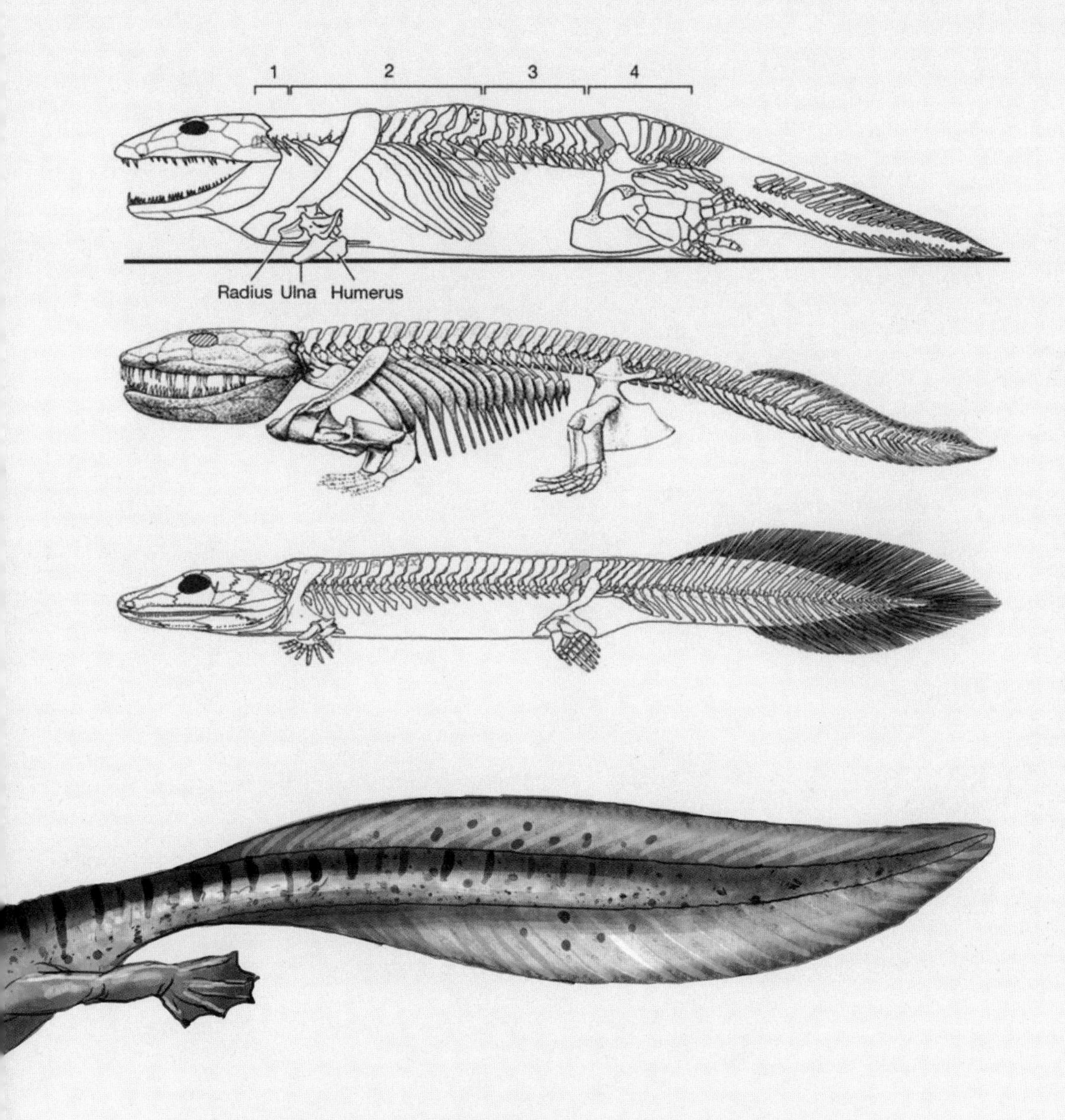


Top: P.E. Ahlberg and others, "The axial skeleton of the Devonian tetrapod *Ichthyostega*," *Nature*, vol. 437, 2005; Reconstructions of *Ichthyostega* (top, middle) and *Acanthostega* (bottom), 1987.

Left and above: Nobu Tamura, Reconstruction of *Crassigyrinus scoticus*, 2007.

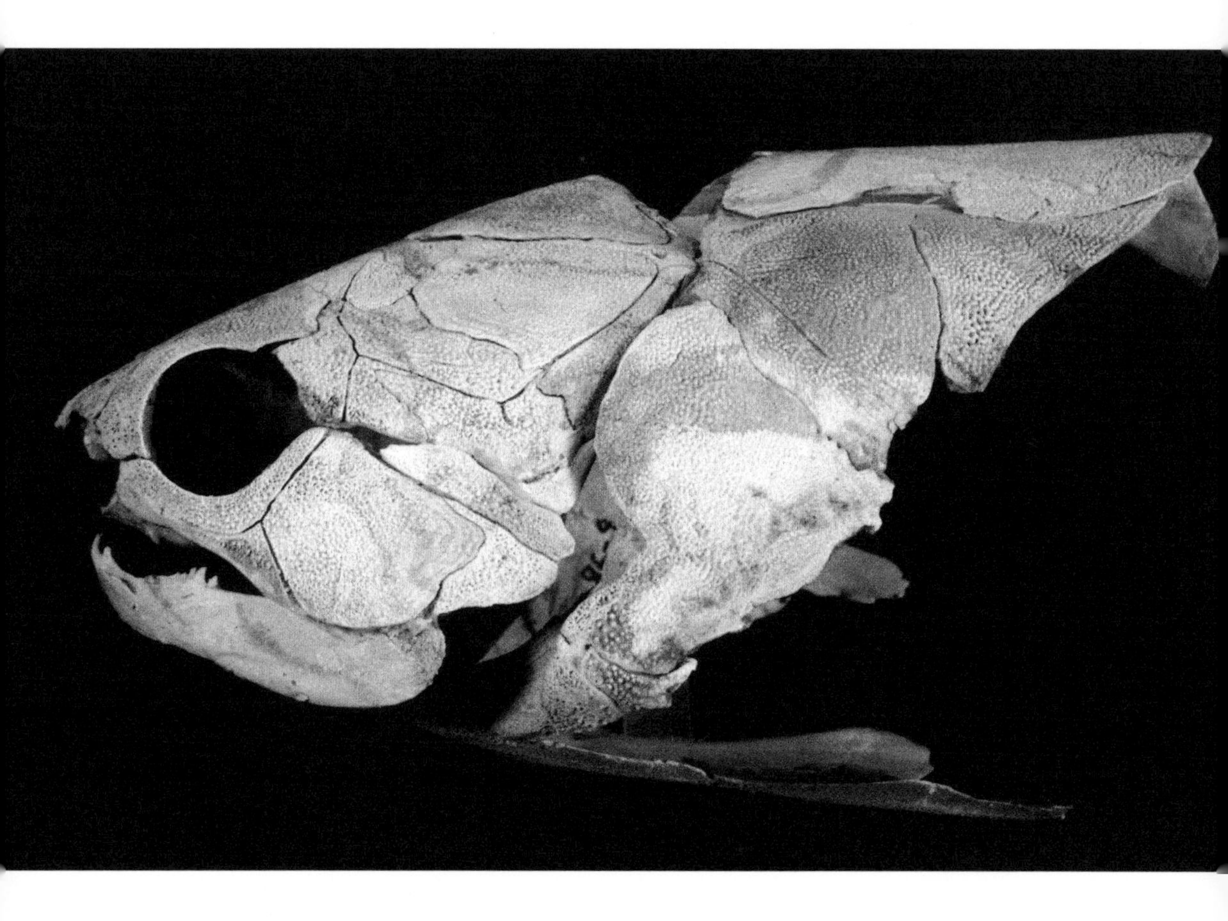

***Mcnamaraspis kaprios*, found by John Long at the Gogo Formation, Canning Basin, Kimberley, Western Australia, in 1986.**

The official state fossil of Western Australia, this placoderm fish *Mcnamaraspis* comes from the late-Devonian, approximately 380-million-year-old Gogo Formation of Western Australia. Australia's outback has yielded myriad placoderm fossils, and those from Gogo can often be separated intact from the surrounding rock by dissolving it away with dilute acid, which sometimes yields three-dimensional specimens.

**Top: Holotype fossil of *Hallucigenia sparsa* from the Burgess Shale. Above: Full reconstruction of *Hallucigenia sparsa* by Jose-Manuel Canete, 2016.**

Stephen Jay Gould's 1989 *Wonderful Life* led to a remarkable popularization of the Burgess Shale fossils (see page 185) and their reappraisal as representing the "Cambrian Explosion": a time of unparalleled morphological experimentation in multicellular animals. Perhaps the most famous of these strange creatures is the aptly named *Hallucigenia*. This creature is now thought to have been widespread, although it is so strange that its relationships to modern animals remain unclear, and it was initially reconstructed back-to-front, upside down, and with the wrong head.

**Above: The Cretaceous dragonfly *Sinaeshcnidia ordonata*, found at the Yixian Foundation, Shangyuan, Liaoning Province, China. Opposite: The Cretaceous bird *Cathayornis yandica*, found at the Jiufotang Formation, Chaoyang, Liaoning Province, China, in 1990.**

If the early nineteenth century was the English era of fossil hunting, and the early twentieth the American, then the early twenty-first may be remembered as the Chinese era. During the Cretaceous Period in what is now Liaoning Province, a series of factors contributed to the deposition of innumerable spectacular fossils, creating one of several important new Lagerstätte. The countryside of that time was a patchwork of lakes and forest welcoming to a diverse fauna, but occasionally engulfed by lethal outgassing from nearby volcanoes. One can easily imagine "Cathay-bird" and "Chinese-dragonfly" tumbling to the ground to be covered by a layer of entombing ash.

**Sue Hendrickson with part of the *Tyrannosaurus rex* fossil "Sue" that she discovered near Faith, South Dakota, in 1990.**

The largest *Tyrannosaurus* skeleton ever discovered, the remarkably intact "Sue" was discovered by, and named after, the fossil hunter Sue Hendrickson. Patterns of bone growth suggest that the animal lived to the age of twenty-eight and died 67 million years ago. Almost as soon as the specimen was discovered a dispute erupted about its ownership, leading to an FBI raid, confiscation of the bones, and the eventual imprisonment of the president of the Black Hills Institute. The skeleton was subsequently auctioned, and bought for more than eight million dollars by the Field Museum, with help from Disney and McDonald's, among other donors.

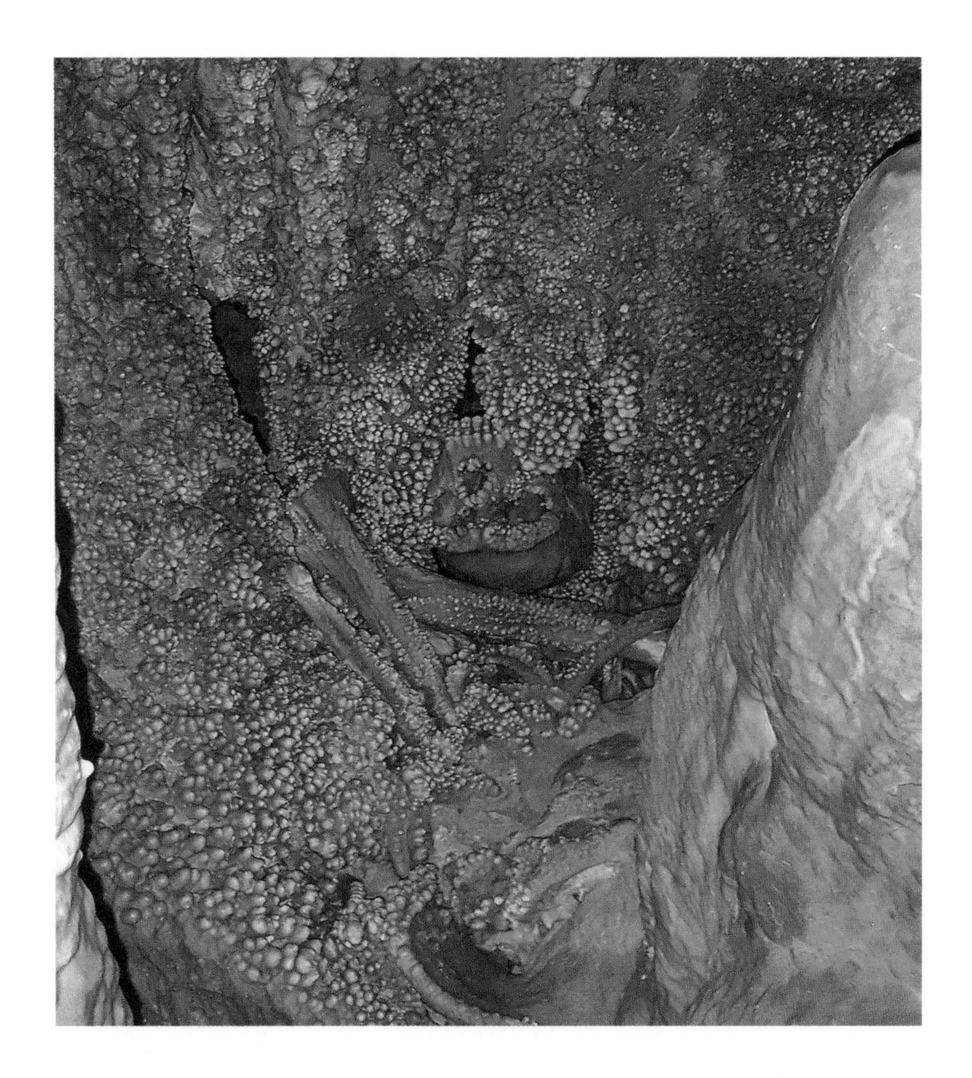

**The Pleistocene Epoch Neanderthal fossil "Altamura Man" (surrounded by calcite concretions), found in Lamalunga Cave, near Altamura, Italy, in 1993.**

"Altamura Man" is the fossil of a male Neanderthal who became lost and perished in a cave in southern Italy between 170 and 130 thousand years ago (0.17–0.13 million years). His skeleton was slowly encased in calcite concretions, a process which stabilized the bones, rendering this one of the most complete Neanderthal fossils ever found. Discovered in 1993, at the time of writing Altamura Man remains lodged in his cave twenty minutes' clamber from the surface, although he has been heroically radiographed in situ and has provided our oldest sample of Neanderthal DNA. It is not known when the lineages of *Homo neanderthalensis* and *Homo sapiens* split, but Neanderthals lived across Eurasia between 400,000 and 40,000 years ago, coexisting with *sapiens* for many millennia.

# KING KONG (1933) AND JURASSIC PARK (1993)

## Paleontology and Popcorn

**Opposite: Models of King Kong and *Stegosaurus* created by John Cerisoli for the 1933 movie *King Kong*.**

**Below: Grant runs from the dinosaur stampede in the 1993 movie *Jurassic Park*, directed by Steven Spielberg, U.S., Universal Pictures.**

It is no coincidence that gigantic beasts are at the core of two of the movies which most stunned contemporary audiences. The original pre-Hays-code *King Kong* remains shocking even today, with its sense of looming dread of the ferocious unknown, and its overt and sometimes bestial sensuality. The stop-motion pounding violence of Kong's battles with the dinosaurs is a key element in building his status as King of the Beasts, in a set-piece which eclipses even his last stand on the Empire State Building.

And sixty years later, the first *Jurassic Park* film was similarly alien to its audience. Cleverly mixing puppetry, computer-generated images, and the director's mastery of the fear of the unseen, the film was the first time screen dinosaurs looked *real*. The *Gallimimus* stampede scene, shown below, complete with deliberate motion blur, is the moment when the Park comes closest to engulfing its visitors, and even the cinema audience.

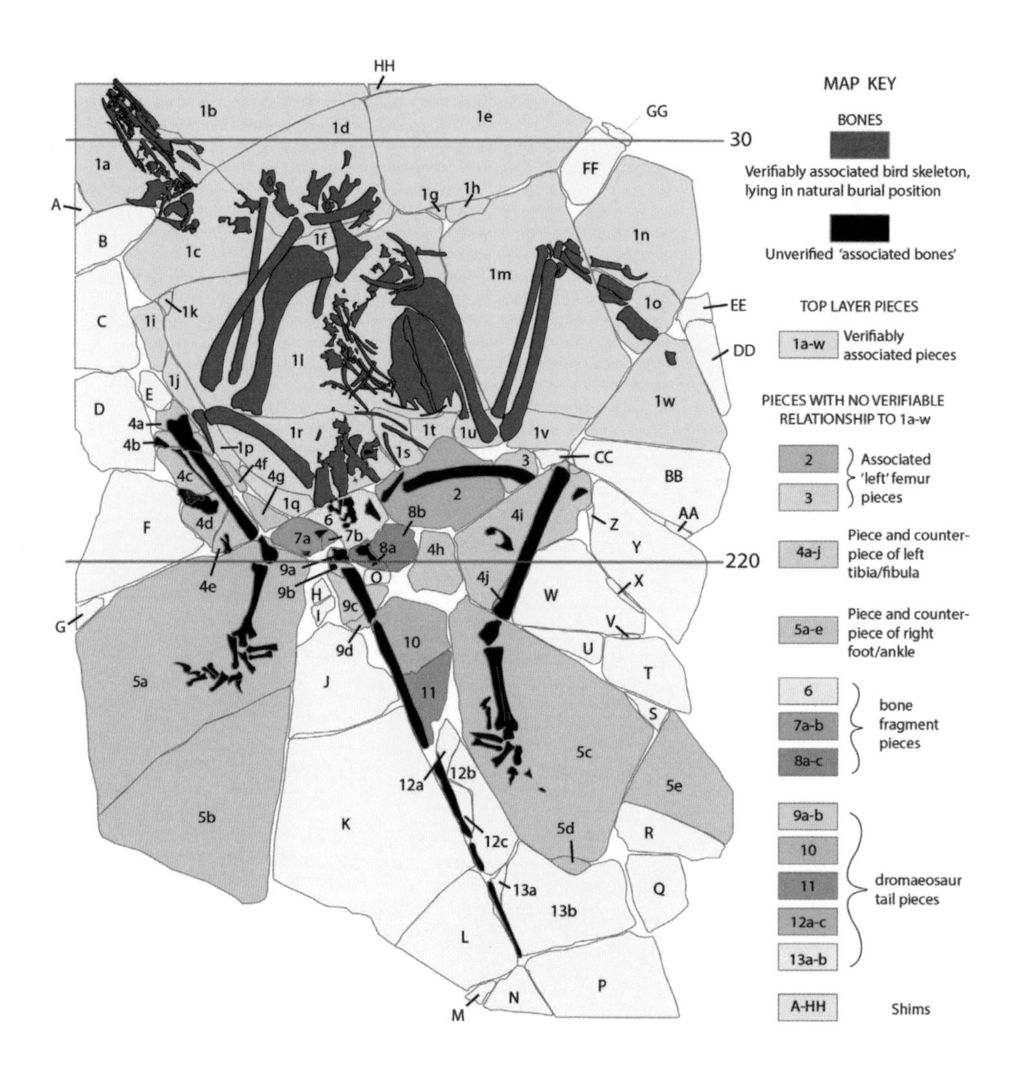


**Timothy Rowe and others, "X-ray computed tomography datasets for forensic analysis of vertebrate fossils," *Nature Scientific Data*, vol. 3, 2016; Map of the "Archaeoraptor" amalgamation, as it was presented for CT scanning at the University of Texas.**

In 1999 a new fossil was announced to the world which was claimed to be an important "missing link" between bipedal theropod dinosaurs and birds. However, the provenance of "Archaeornis" attracted immediate suspicion. The fossil is made up of tens of fragments cemented onto a backing slab, and had been smuggled out of China to the US to be auctioned commercially before it was eventually repatriated. Then, a 2001 report appeared in the distinguished journal *Nature* simply titled "The Archaeoraptor Forgery." Its authors used computed tomography (CT) to scan the fossil and showed that it had been fabricated from fragments of two fossils—one a bird and one a dromaeosaurid dinosaur. In addition, fossils often emerge from rocks as a pair—the fossil itself and a counterpart imprint on the adjacent layer of stone—and Archaeoraptor's "two feet" were actually two such mirror-image imprints of a single foot. Worst of all, the forgers clearly damaged two valuable fossils to create what has now been dubbed "Piltdown chicken."

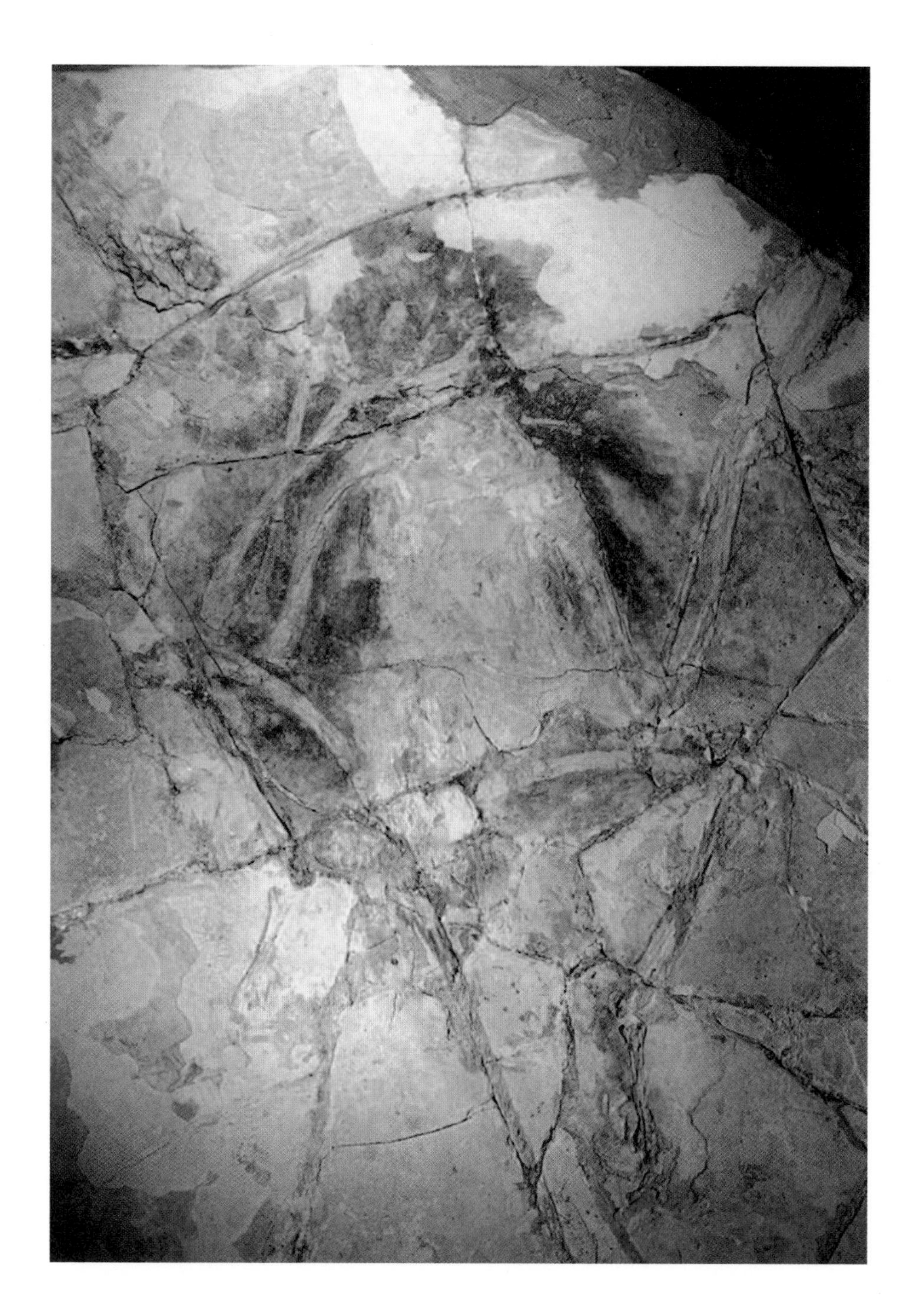

Photograph of the "Archaeoraptor" amalgamation taken as it was presented for CT scanning at the University of Texas in 1999.

**The Cretaceous Period ankylosaur *Liaoningosaurus paradoxus* found at the Yixian Formation of Liaoning Province, China, in 2001.**

A unique fossil from China, *Liaoningosaurus* overturned the idea that ankylosaurs were all huge spiky dinosaurs like those depicted fighting *Tyrannosaurus* in children's picture books. Ankylosaurs belong to the half of dinosaurs classified as "bird-hipped" ("ornithischian," as opposed to lizard-hipped "saurischian" forms) and *Liaoningosaurus* is the only known flesh-eating member of that group. Discovered in 2001, it was found with fish bones in its gut and possesses fish-piercing spikes on its teeth. It also has the distinction of being the smallest known ornithischian dinosaur and the only armored aquatic dinosaur—although its armor lies mainly on its underside, so presumably this little piscivorous creature's main threats came from the deep.

**Reconstruction by Fabrizio de Rossi of the fish-eating ankylosaur *Liaoningosaurus paradoxus*.**

**Skeleton of a pregnant ichthyosaur (the Jurassic *Stenopterygius quadriscissus*), with three embyros inside and another in the birth canal, found near Holzmaden, Germany, in 1990.**

All present-day tetrapods may be divided into "lissamphibians" (including frogs and salamanders) which must return to water to lay their eggs, and "amniotes" (including reptiles, birds, and mammals) which, if they lay eggs, must lay them on land. Because of this, any amniote reverting to an aquatic existence, to the extent that it can no longer clamber up onto land to lay eggs, must give birth to live young. And indeed there is evidence that the Mesozoic marine ichthyosaurs (see page 52) gestated their offspring. In some fossils multiple embryos can be seen within an adult ichthyosaur's body cavity, and sometimes even emerging from it.

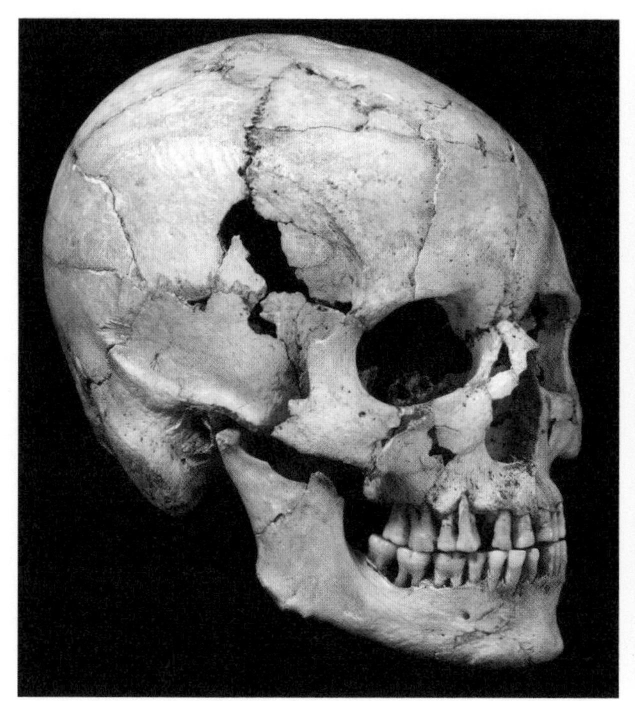

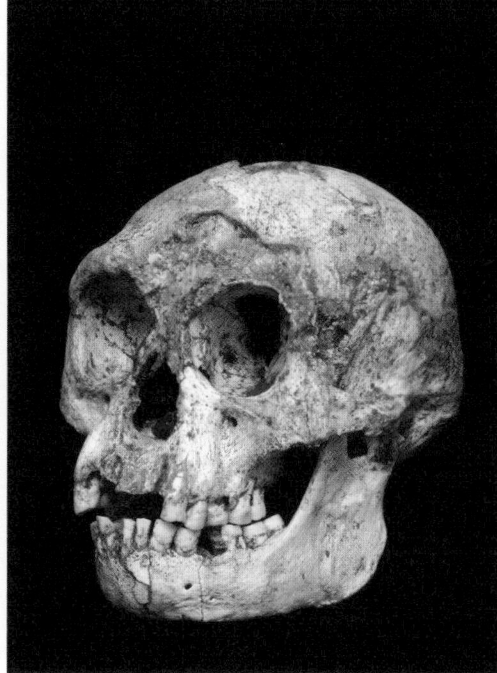

**Skulls of *Homo sapiens* (right) and *Homo floresiensis* (left), the latter found in the Liang Bua limestone caves, Indonesia, in 2003.**

The 2003 discovery of fossils of diminutive hominins on the Indonesian island of Flores created headlines worldwide. These *Homo floresiensis* ("Flores Man" or "hobbit") individuals—of which nine have now been found—were only 3½ft (1m) tall and it was initially thought that they had survived until 12,000 years ago. We now think the fossil and artefact evidence indicates a tenure on the island from at least 200,000 to 50,000 years ago (0.20–0.05 million years). The most confusing feature of *Homo floresiensis* is that its brain was similar in size to that of a chimpanzee—a third the size of modern humans—yet there is evidence these hominins fashioned stone tools like other, larger-brained types. So many theories of human evolution are based on the premise that hominins' large brains allowed them to achieve new feats, that it is confusing such a small-brained hominin acquired these "human-like" skills. However, we still know very little about *Homo floresiensis*' behavior or its relationships to our own species. Rather than being a diminutive offshoot of *Homo sapiens*, or even a diseased colony as some have suggested, it seems most likely "Flores Man" represents the remarkable survival of an ancient hominin lineage dating from the early diversification of *Homo* species.

**The Devonian Period lobe-finned fish *Tiktaalik roseae*, found on Ellesmere Island, Nunavut, Canada, in 2004.**

Discovered in 2004, *Tiktaalik* added another strand to the story of vertebrates' transition to land. It is too simplistic to say that *Tiktaalik* was "fishier" than its neighbors *Acanthostega* and *Ichthyostega* (see pages 194 and 147), but it pre-dates them at 375 million years ago and its "hands" and "feet" were probably more fin-like. *Tiktaalik* was large—around 10ft (3m)—and probably had gills, but it also had lungs and a mobile, tetrapod-like neck with a flattened crocodile-like head. In addition, the hind limb is as large as the forelimb—previously thought to be a feature of "four wheel drive" land vertebrates rather than "front wheel drive" fish.

**C. Kevin Boyce and others, "Devonian landscape heterogeneity recorded by a giant fungus," *Geology*, vol. 35, 2007; *Prototaxites* fossil in situ.**

The paleontology of fungi has been less studied than that of plants and mammals, but *Prototaxites*' size meant it was noticed as early as 1843. Its huge trunks are up to 30ft (9m) in length, and they were for some time misidentified as conifers—yet these fungi are too ancient, at 470–360 million years old, for this to be true. It was, in fact, the largest known terrestrial organism of its time, although it is not known how such a large fungus gained its sustenance.

# YURI KHUDI (BIRTH DATE UNKNOWN, DISCOVERY IN 2007)

## Reawakening *Mammuthus primigenius*

Locals have always known that mammoths once lived in Siberia. When spring brings the thaw, or when meandering rivers gouge into the soil, tusks fall out—often leading to elaborate and sometimes dark legends about the previous inhabitants of this remote region.

Yet in 2007 came a particularly impressive discovery, when the reindeer herder Yuri Khudi and his children discovered an unusually complete and well-preserved baby woolly mammoth, *Mammuthus primigenius*, frozen solid by the side of the Yuribey River. The individual was amazingly intact—it had lost its toenails and hair, and some local dogs were to gnaw on its ear, but otherwise it appeared perfect—even its eyelashes were

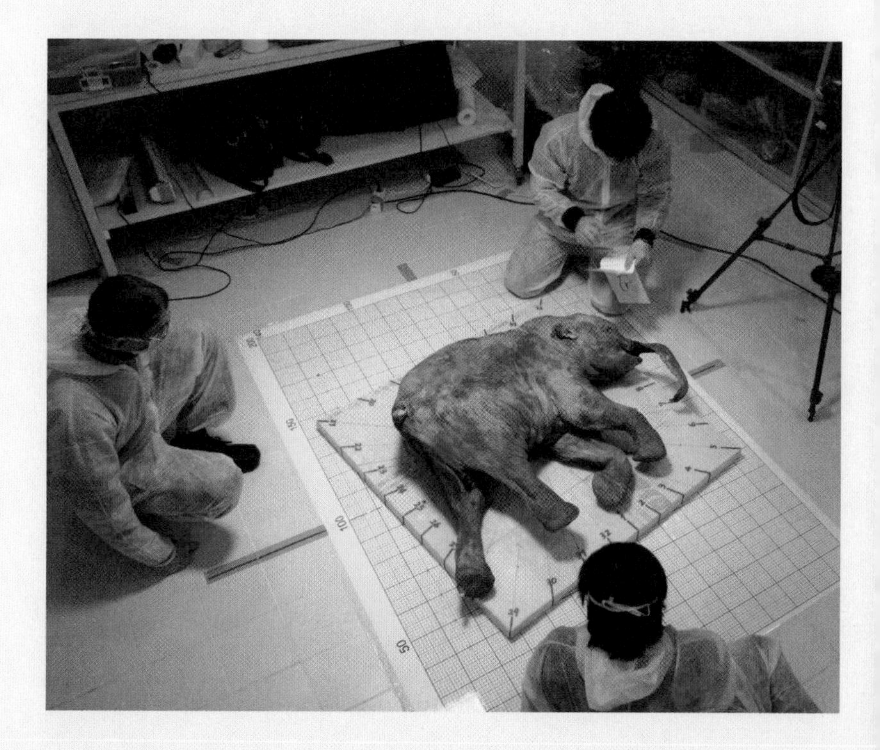

**"Lyuba" being examined by scientists at the Shemanovskii Regional Museum, Salekhard, Russia, in 2008.**

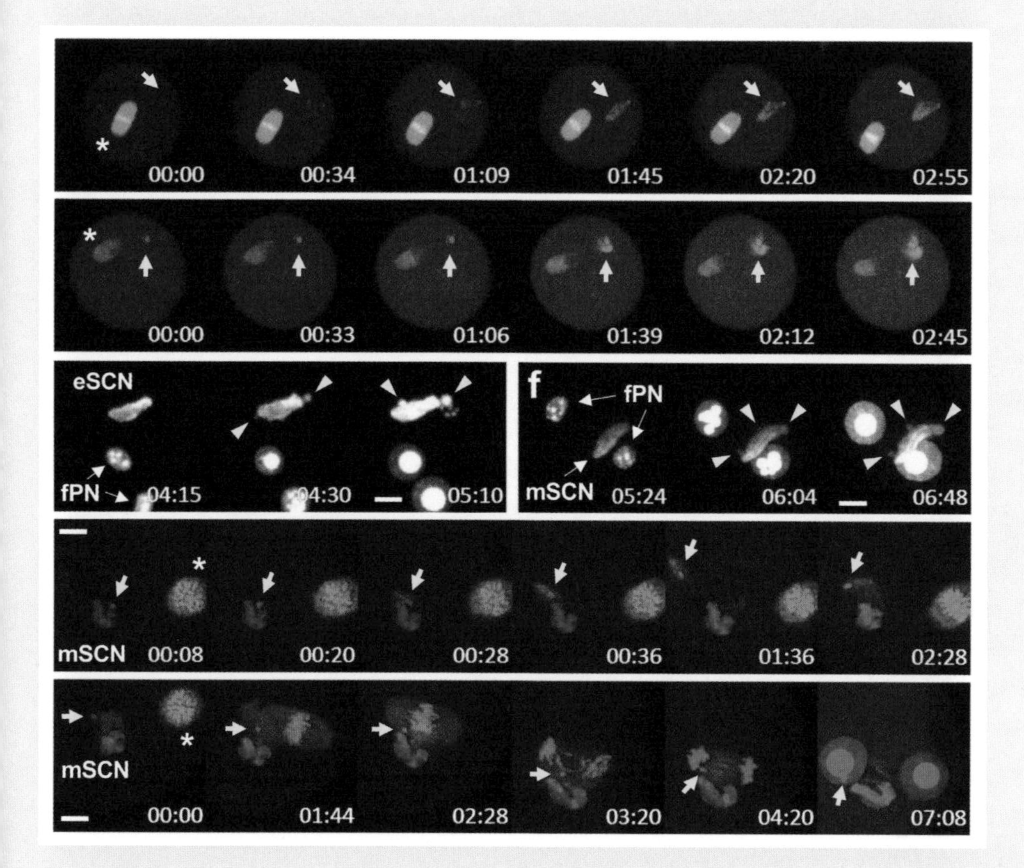


intact. The mammoth was soon informally named after Yuri's wife Lyuba, and estimates of its antiquity now range between 28,000 and 43,000 years (0.028–0.043 million years).

Understandably, "Lyuba" has been the subject of considerable study—she has been CT scanned and her DNA sampled. One of the most dramatic studies was published in 2019, when groups at several institutions in Japan attempted to reactivate Lyuba's genetic material. Going well beyond simply extracting and characterizing her DNA, they harvested cell nuclei from relatively undamaged Lyuba cells and transplanted them into live mouse egg cells. Remarkably, the transferred material showed signs of activity: genetic material was either transported toward the mouse DNA, or formed structures akin to the "pronuclei" which appear just before a fertilized egg starts to divide—although the merged cells did not actually replicate. The cellular machinery involved in these processes is extremely complex, so for it to retain any function after tens of thousands of years in the Siberian ice is astounding.

**Kazuo Yamagata and others, "Signs of biological activities of 28,000-year-old mammoth nuclei in mouse oocytes visualized by live-cell imaging,"** ***Scientific Reports,*** **vol. 9, 2019; Reconstruction of mammoth nuclei upon nuclear transfer to mouse oocytes.**

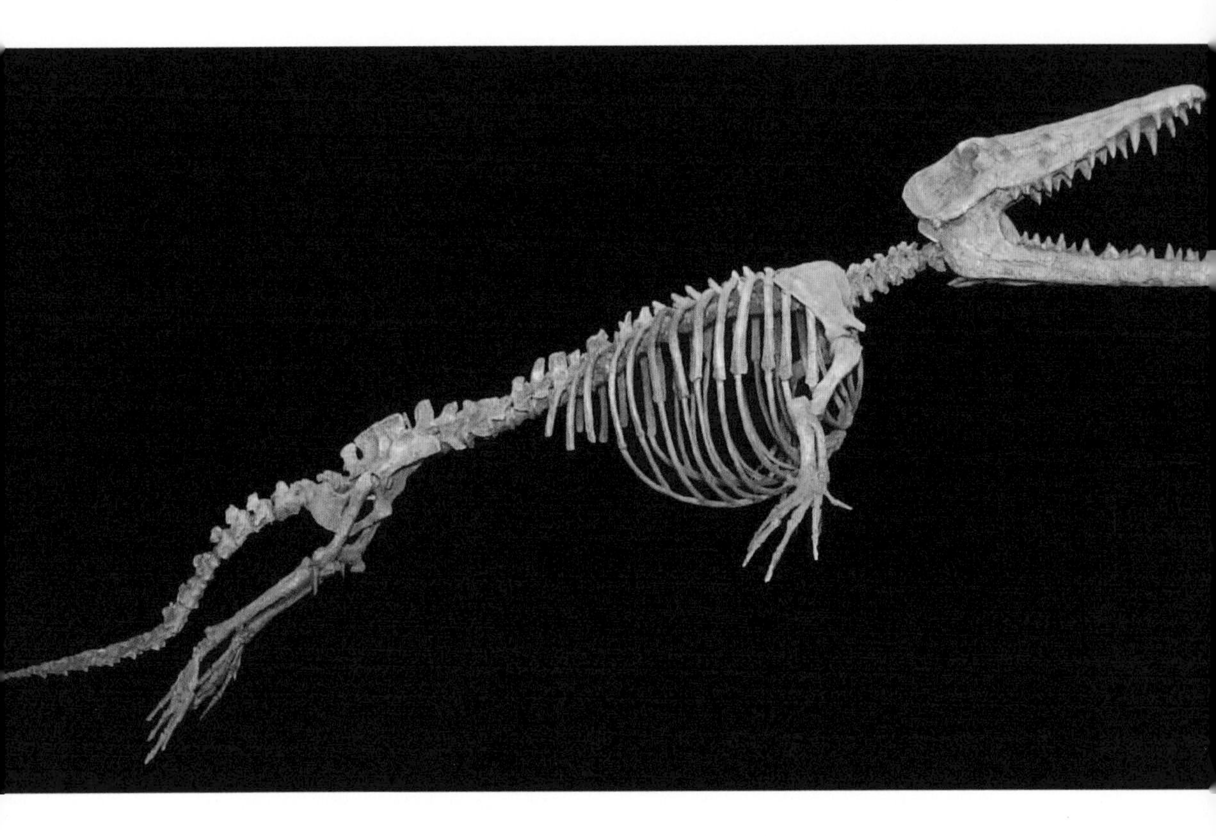

**Hans Thewissen, "From Land to Water: the Origin of Whales, Dolphins, and Porpoises,"** ***Evolution: Education and Outreach*****, vol. 2, 2009; The Eocene Epoch cetacean "missing link"** ***Ambulocetus natans*****.**

Mammals such as seals and sea lions, manatees, otters, and cetaceans have reverted to a marine existence on several occasions—and many convincing transitional fossil forms have been found. *Puijila darwini* is the oldest known fossil near the evolutionary path leading to seals and sea lions, or "pinnipeds." Its skull has some very seal-like features, but it retains limbs and paws and it probably had an otter-like appearance. Indeed, pinnipeds are thought to be members of the Carnivora group, which also includes otters, polecats, skunks, cats, dogs, and bears. In contrast, the cetaceans are descended from hippopotamus-like hoofed mammals. *Ambulocetus natans* (swimming-walking-whale), which was discovered in Pakistan, is a close relative of those limbed ancestors of modern whales and dolphins.

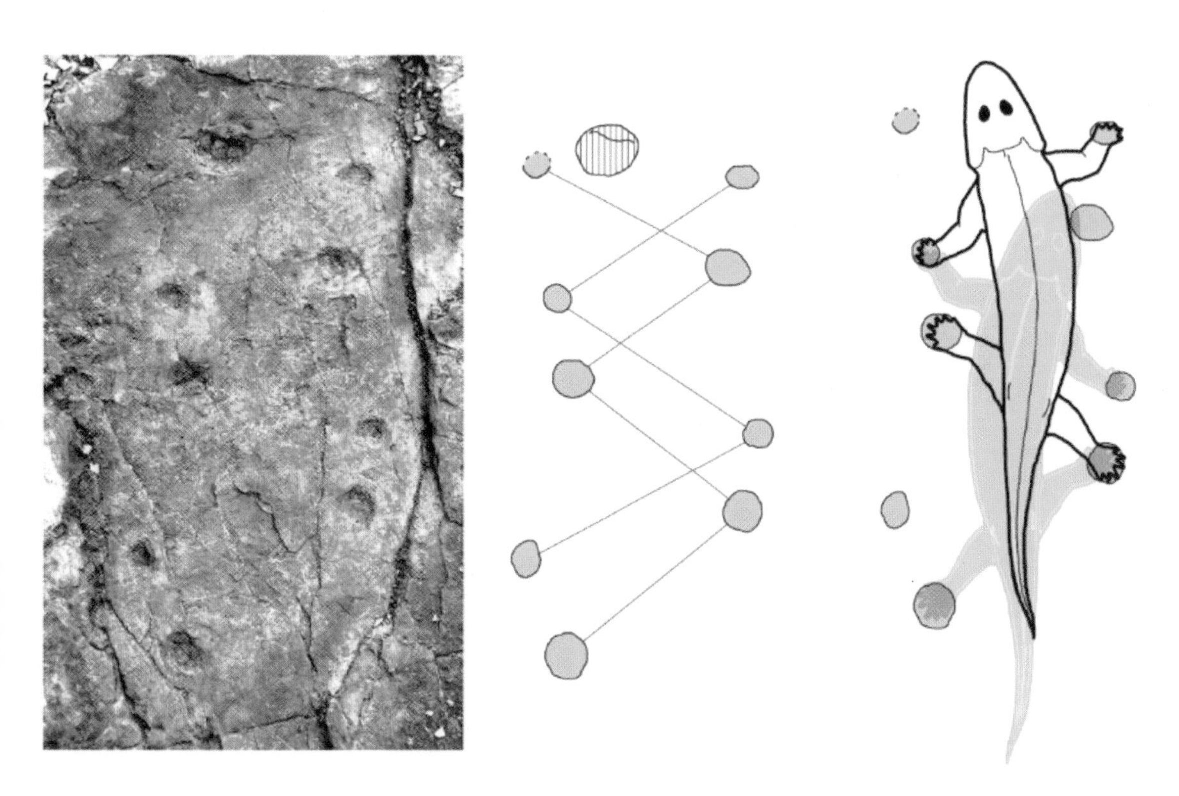

**Grzegorz Niedźwiedzki and others, "Tetrapod trackways from the early Middle Devonian period of Poland," *Nature*, vol. 463, 2010.**

A fossil that paints an evocative picture of ancient life, this is a trackway from approximately 395 million years ago, made by an early tetrapod (land vertebrate) as it plodded across a marine tidal flat in what is now Poland. These tracks precede the oldest known tetrapod fossils by 18 million years, and suggest that tetrapods actually coexisted with their finned "elpistostegid" close relatives for millions of years.

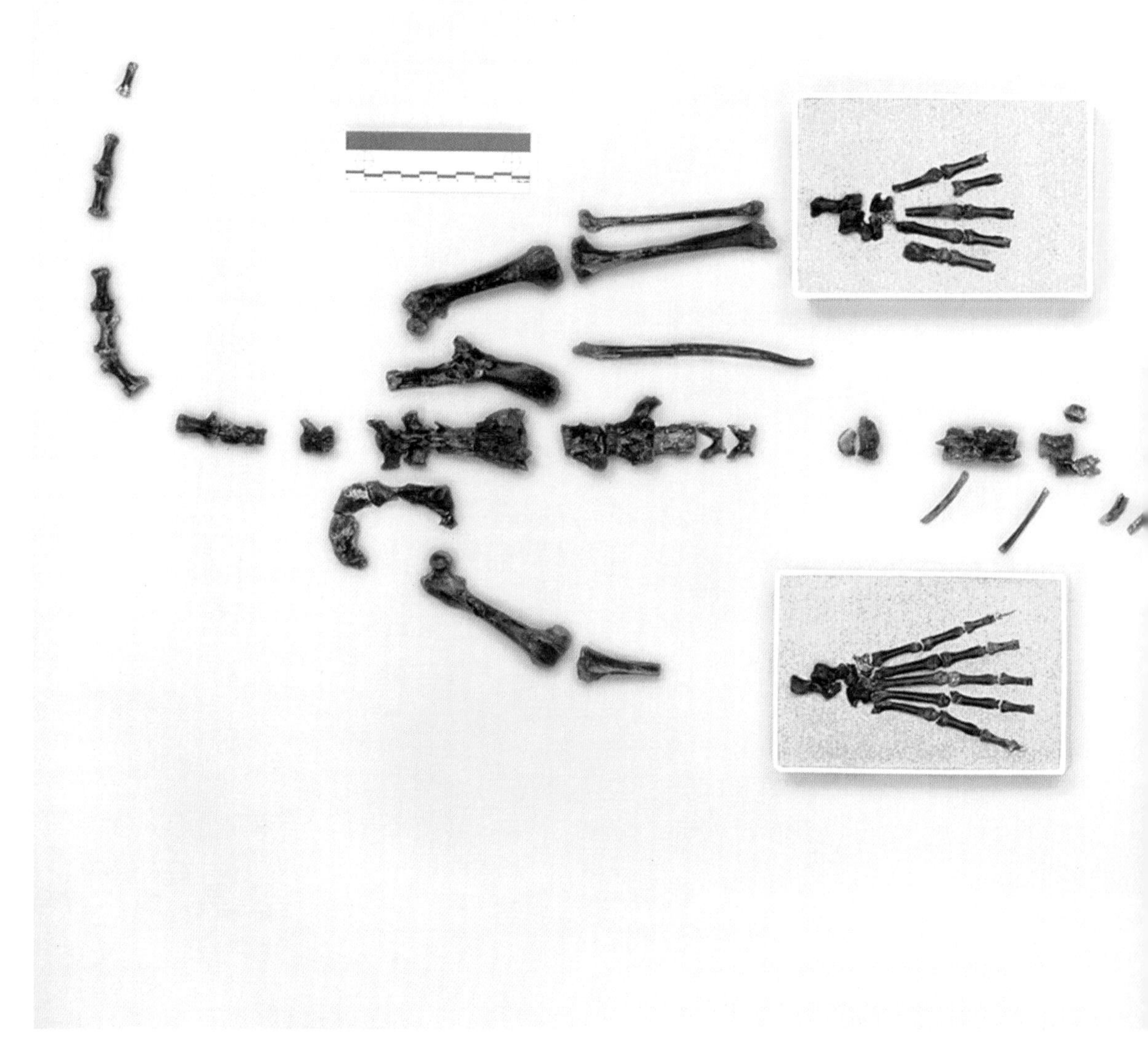

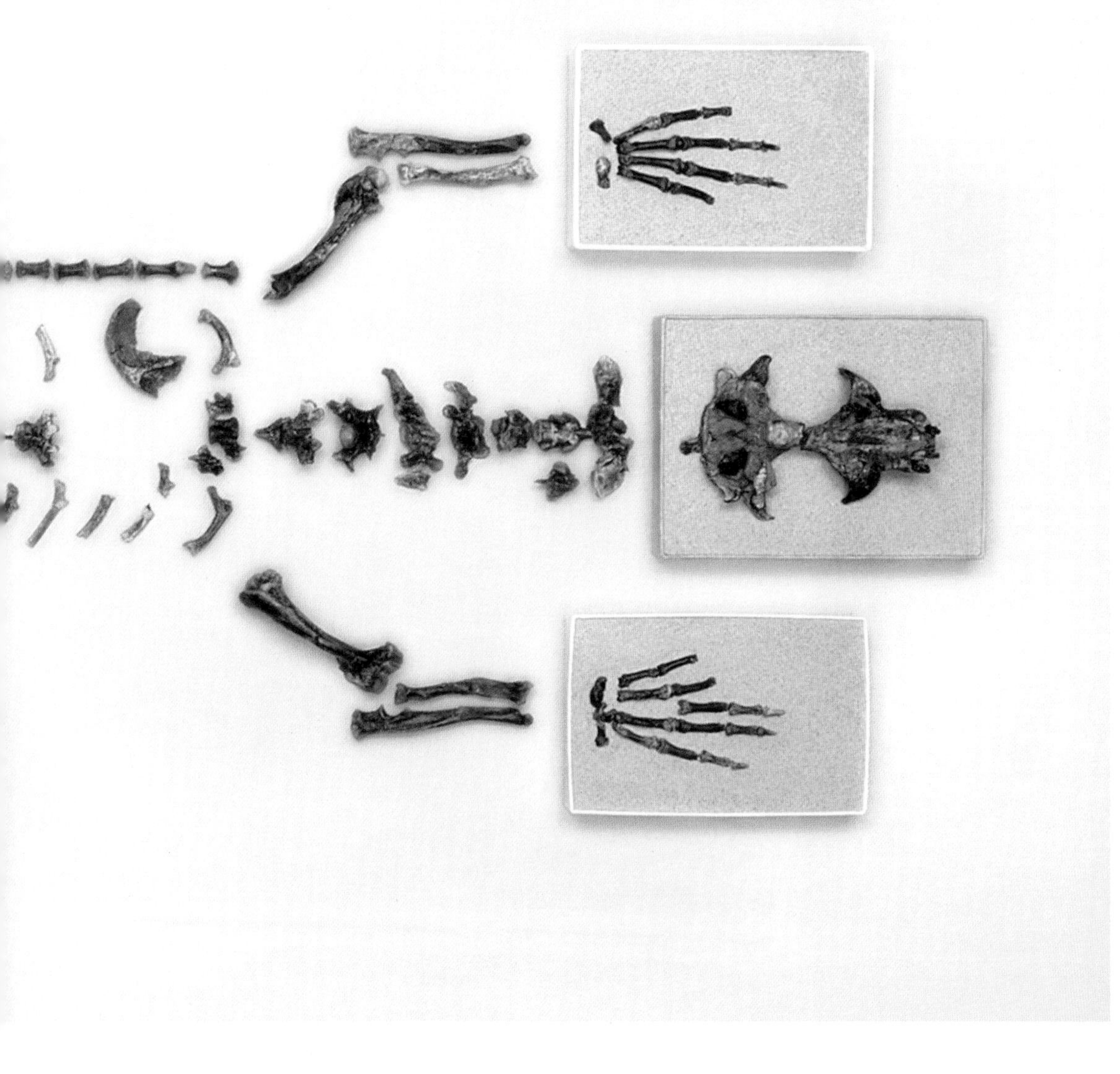

**The Oligocene-Miocene Epoch "missing link" seal *Puijila darwini*, discovered on Devon Island, Canada, in 2007.**

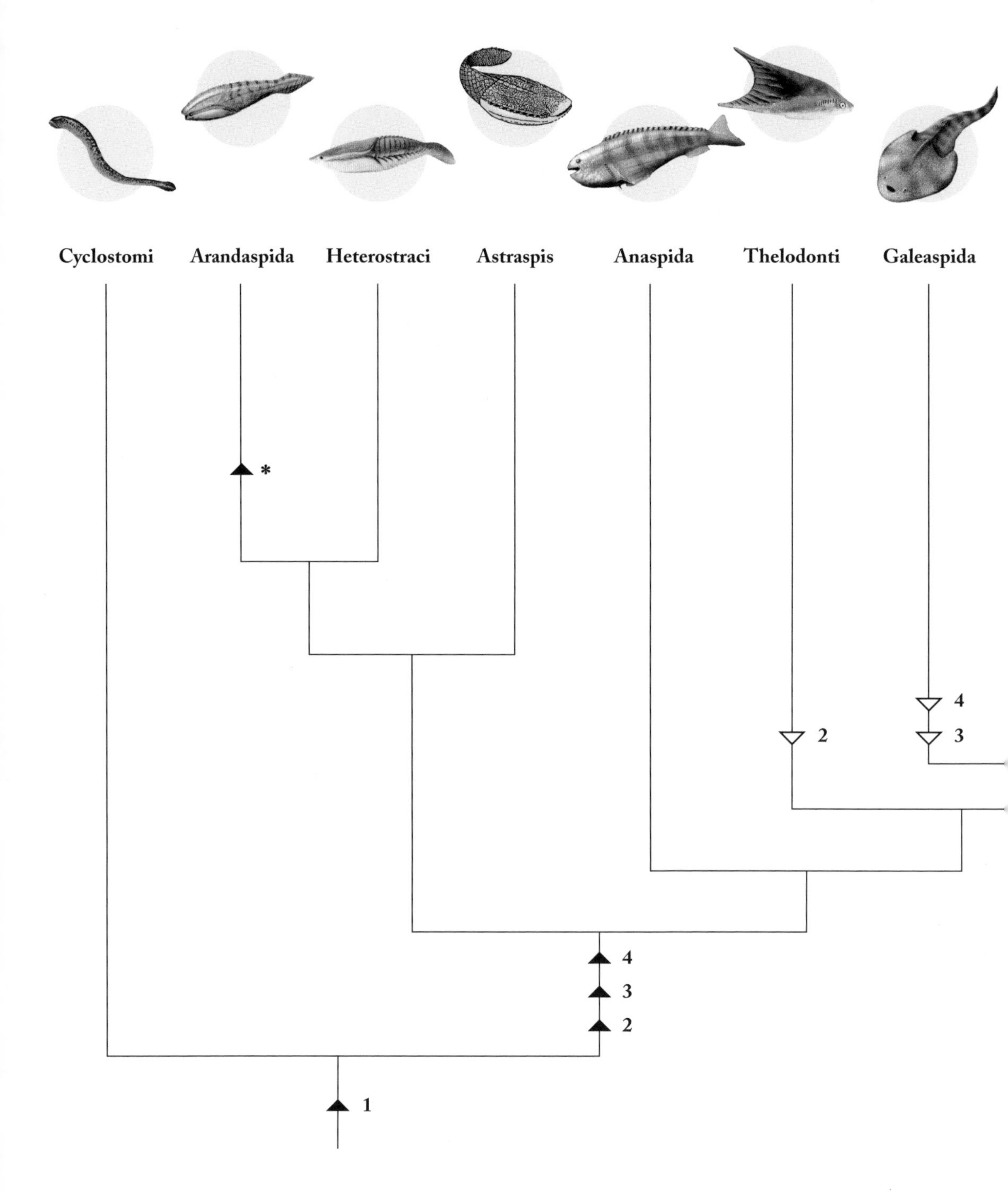
Cyclostomi
Arandaspida
Heterostraci
Astraspis
Anaspida
Thelodonti
Galeaspida
*
4
2
3
4
3
2
1

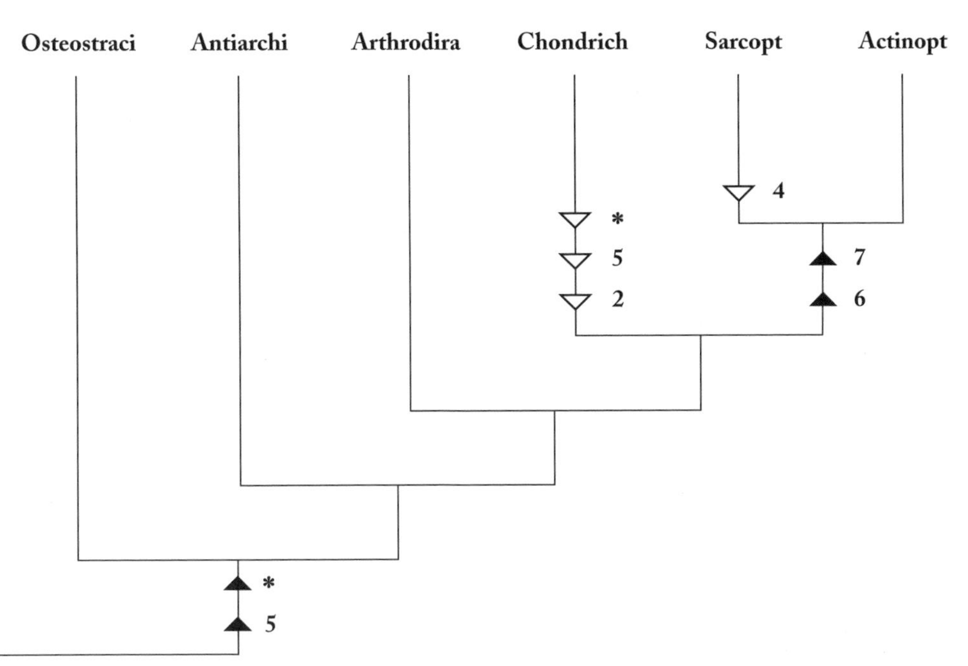


**Relationships between vertebrate groups, adapted from Joseph Keating and others, "The nature of aspidin and the evolutionary origin of bone," *Nature Ecology & Evolution*, vol. 2, 2018.**

Fish were the first vertebrates, and most vertebrates are fish, so determining their relationships lies at the core of vertebrate biology—continuing a tradition started by Louis Agassiz in the nineteenth century (see page 51). This proposed classification appears in an analysis of the all-important evolution of vertebrate skeletal tissues. Only four of these groups remain today—on the left, "Cyclostomi" includes the jawless lamprey and hagfish, and on the right, "Chondrich" is the rays and sharks. "Actinopt" is most bony fish and "Sarcopt" is the coelacanth, lungfish, and all land vertebrates.

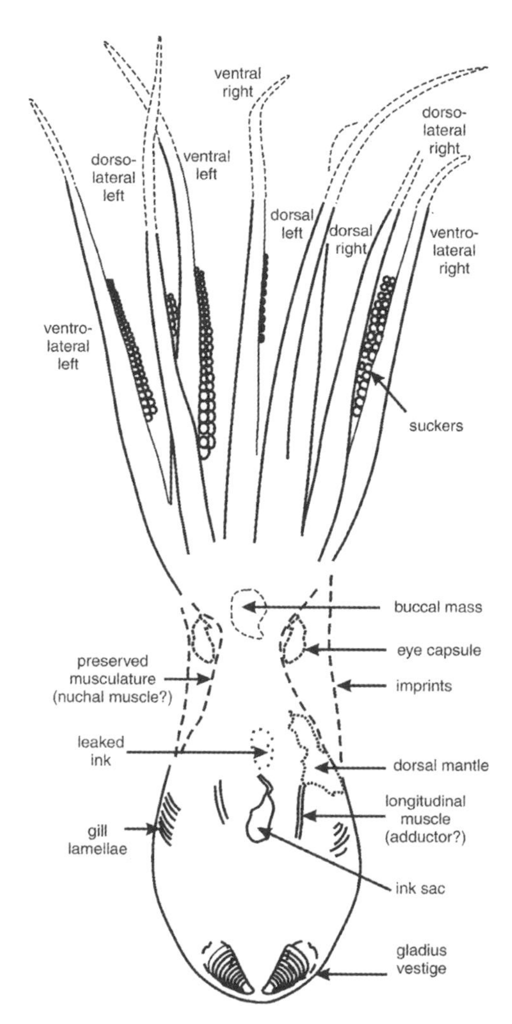


**Dirk Fuchs and others, "New octopods (Cephalopoda: Coleoidea) from the Late Cretaceous (Upper Cenomanian) of Hâkel and Hâdjoula, Lebanon," *Palaeontology*, vol. 52, 2008; *Keuppia levante* from the Upper Cenomanian.**

Wider searches of fossil Lagerstätten have revealed sediments in which the preservation of soft tissues is truly remarkable. This specimen is an easily recognizable octopus from 95-million-year-old rocks in the Lebanon. Some octopuses contain a rigid bar or *gladius* (sword), thought to be a relic of their past as a shelled mollusc.

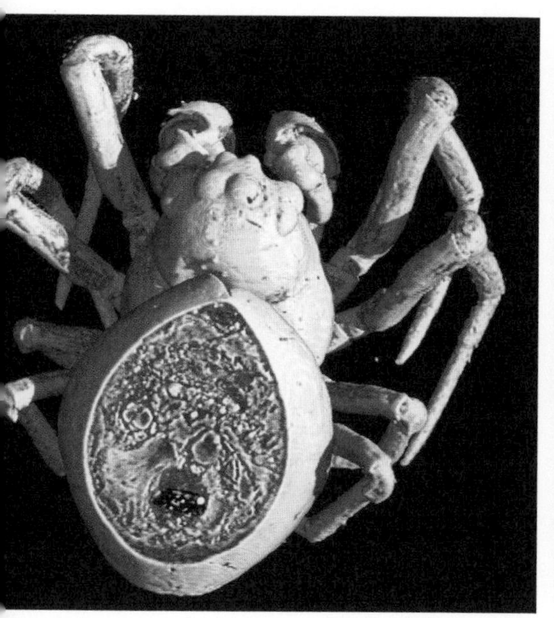

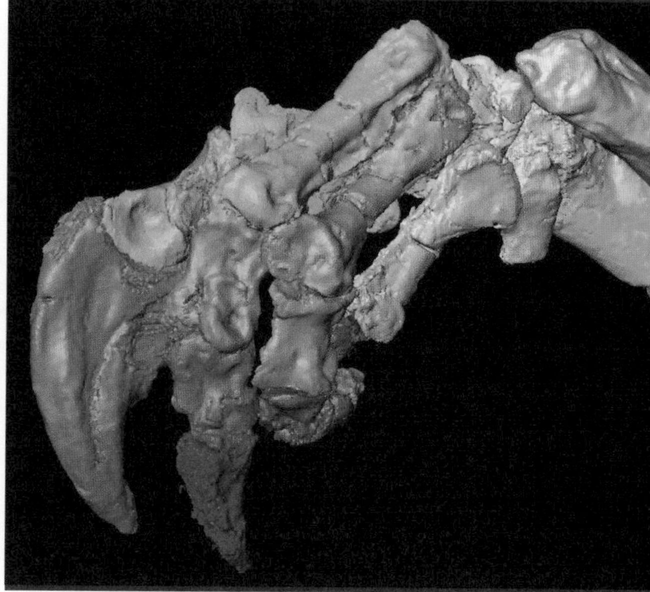

**Above left: Manuel Dierick and others, "Micro-CT of fossils preserved in amber," *Nuclear Instruments and Methods in Physics Research A*, vol. 580, 2007; An approximately 1-mm-long spider belonging to the Micropholcommatidae family. Above right: CT scan by Timothy Rowe and others of the left antebrachium (forearm) and manus (hand) of *Sarahsaurus aurifontanalis*, 2010.**

Computed Tomography, or CT, is a technique that generates three-dimensional X-ray radiographs. Modern CT technology can now produce images of fossil organisms still embedded in rock, and also resolve structures down to the micrometer level (one thousandth of a millimetre). This image shows the external surface of a tiny spider found in amber (fossilized tree resin) and a "slice" through its abdomen. On a different scale, the image opposite shows the "forearm" (yellow) and "hand" (green, blue, purple) of *Sarahsaurus* from the Kayenta formation of the southwestern United States, which is approximately 190 million years old.

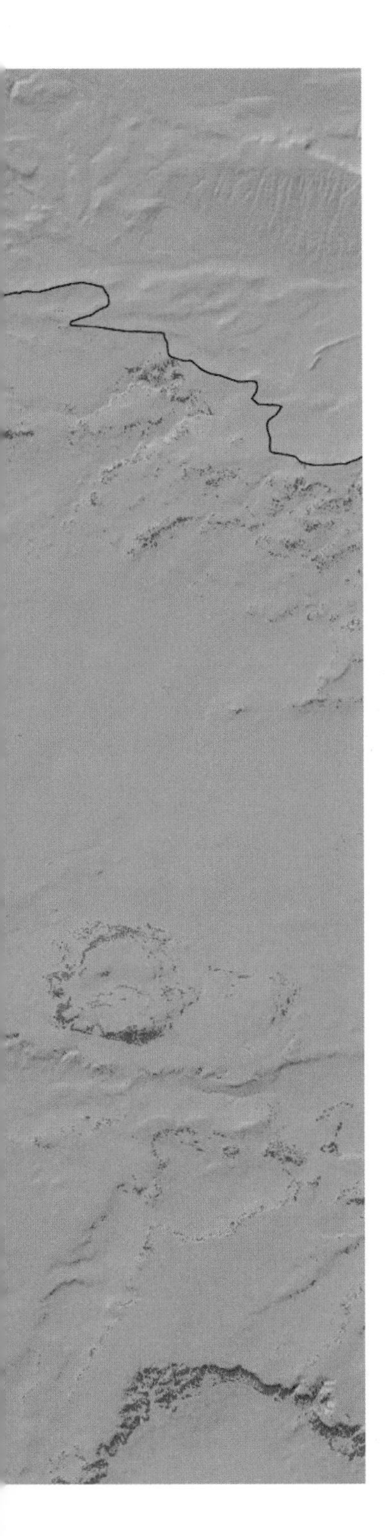

**Potential fossil-bearing locations in the Great Divide Basin (opposite), a newly-discovered fossil location, photographed from ground level (above). Robert Anemone and others, "Finding fossils in new ways: An artificial neural network approach to predicting the location of productive fossil localities,"** ***Evolutionary Anthropology*****, vol. 20, 2011.**

The science of paleontology is obviously dependent on the discovery of fossils, yet fossiliferous locales are often found by simple chance or as the result of a fossil hunter's hunch. This study was an attempt to train a computerized "artificial neural network" to use satellite-derived maps to identify the locale types where fossils had tended to be discovered in the past—so that it could then pinpoint optimal sites for future excavation. Harking back to the golden age of American fossil hunting, this trial focused on Paleocene and Eocene fossil mammals in Wyoming.

**Top: Children at Nanjing Paleontology Museum view exceptionally well-preserved fossil fauna, about 530 million years old, from the Maotianshan Shale deposits in Chengjiang, Yunnan Province, 2006.**

**Above: Reconstruction of *Jianshanopodia decora*, a Cambrian lobopodian discovered in the Chengjiang deposits in 2017.**

Chengjiang is a fossil Lagerstätte in Yunnan, China, which preserves fossils with soft tissue details, despite dating from 520 million years ago, before most Burgess Shale fossils. This strange species, *Jianshanopodia decora*, is a lobopodian—a group argued to represent intermediates between mysterious Cambrian soft-bodied forms and today's arthropods—insects, arachnids, crustaceans, millipedes, and centipedes.

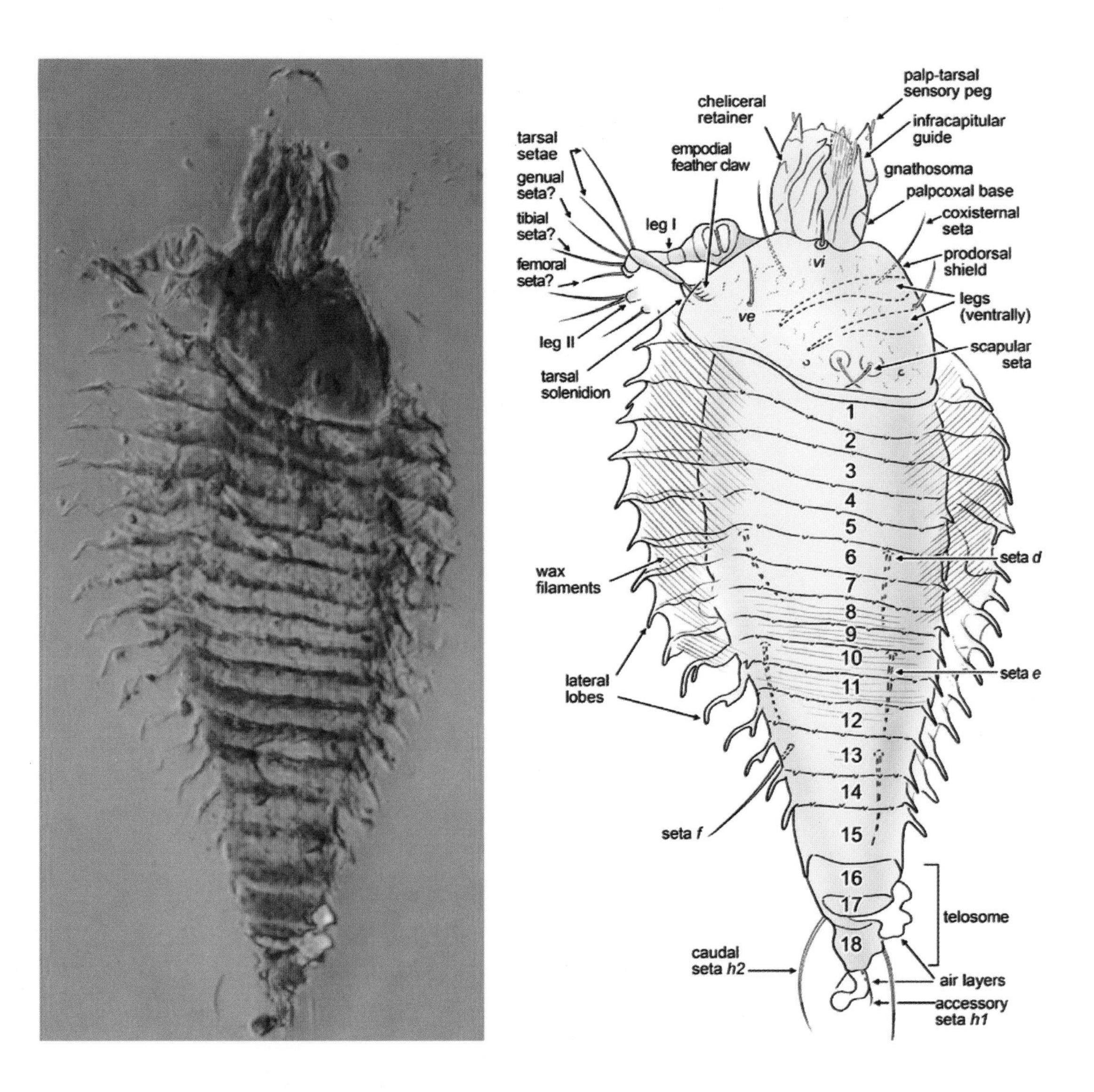


**Alexander Schmidt and others, "Arthropods in amber from the Triassic Period," *Proceedings of the National Academy of Sciences*, vol. 109, 2012; Eriophyoid mite in Italian Triassic amber.**

Tree resin fossilizes into amber when the pressure of overlying sediments, as well as heat, chemically alters it into a robust, long-lived form—although most tree resin degrades long before this can occur. This image of an entombed mite is from a description of exceptionally old arthropods in amber, from approximately 230 million years ago.

# LIFE ON EARTH (4,500,000,000 YEARS AGO TO THE PRESENT)

## Fossils From When the World Was Young

Little is known about the origin of life on Earth. The evolution (if that is the correct term) from non-biological material, of enclosed cell-like structures able to metabolize nutrients and self-replicate, seems to require such a remarkable coincidence of conditions, events, and time, that scientists still have little idea how it happened. However, one thing is clear—the further back in Earth's 4,540-million-year history one looks, the further back life seems to appear.

For much of the twentieth century, Walcott's Burgess Shale fossils were the oldest known, at around 509 million years (see page 154), only superseded when the Ediacaran fauna was characterized as dating

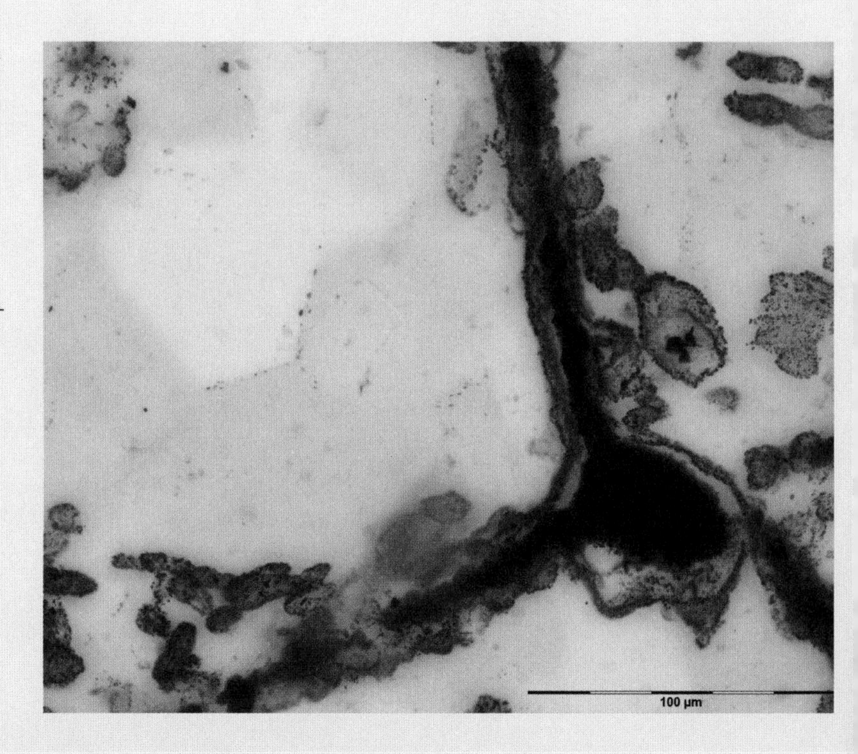


**Matthew Dodd and Dominic Papineau, Microbial haematite filaments attached to a clump of iron in seafloor hydrothermal vent deposits, Nuvvuagittuq Supracrustal belt, Quebec, approximately 3,770–4,280 million years old.**

**Section of a fossilized Cretaceous Period stromatolite slab, found at the El Molina Formation, Bolivia.**

from 600 to 540 million years ago. In fact, Walcott also discovered stromatolites—large fossilized mats of layered cyanobacteria which still grow in places around the world today. Stromatolites are the dominant fossils from 1,200 million years ago, but it now seems some may be as old as a remarkable 3,500 million years.

Cyanobacteria are single-celled photosynthetic bacteria, lacking a nucleus (prokaryotes), which were responsible for the progressive oxygenation of the Earth's atmosphere—slowly from 2,000 million years ago, and rapidly from 1,000 million years ago. The microfossil beds of the Hutuo Group of north China have provided evidence for a diverse array of cyanobacteria dating from 2,150 to 1,950 million years ago, as well as tentative evidence of the earliest nucleated cells—these "eukaryotes" probably evolved when one prokaryote host engulfed other prokaryotes to form a commune which could carry out a wide array of functions. One lineage of eukaryotes even engulfed a cyanobacterium to exploit its photosynthetic abilities, and thus plants appeared.

And remarkably, the story of life now goes back almost as far as it can possibly go. The Nuvvuagittuq Belt in Quebec is a fragment of ancient seabed dating back 3,770 to 4,280 million years, and contains micrometer-scale tubes and filaments of the ferrous mineral haematite—similar to those produced by today's iron-oxidizing submarine hydrothermal vent bacteria. In addition, actual microfossils now date almost as far back, so it seems life appeared surprisingly soon after the planet's oceans formed 4,500 million years ago.

Thus, the story of life on Earth can now be seen as a long phase of microbial diversification, followed by a short period of multicellular complexity, and preceded by an even shorter period of "abiogenesis"—the mysterious commencement of life from non-living material.

**Leiming Yin and others, "Microfossils from the Paleoproterozoic Hutuo Group, Shanxi, North China: Early evidence for eukaryotic metabolism," *Precambrian Research*, vol. 342, 2020; Holotype of *Eoentophysalis hutuoensis*.**

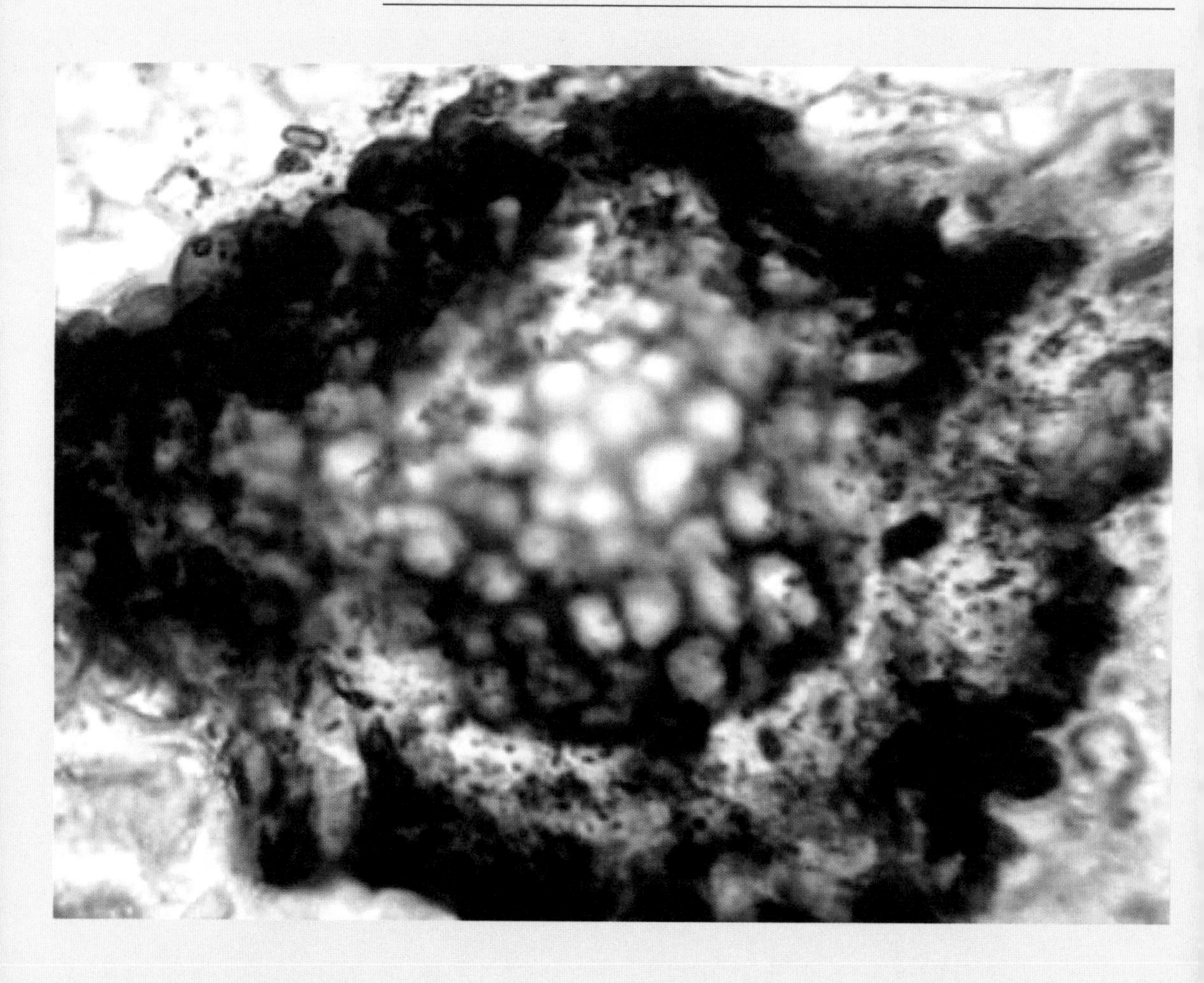

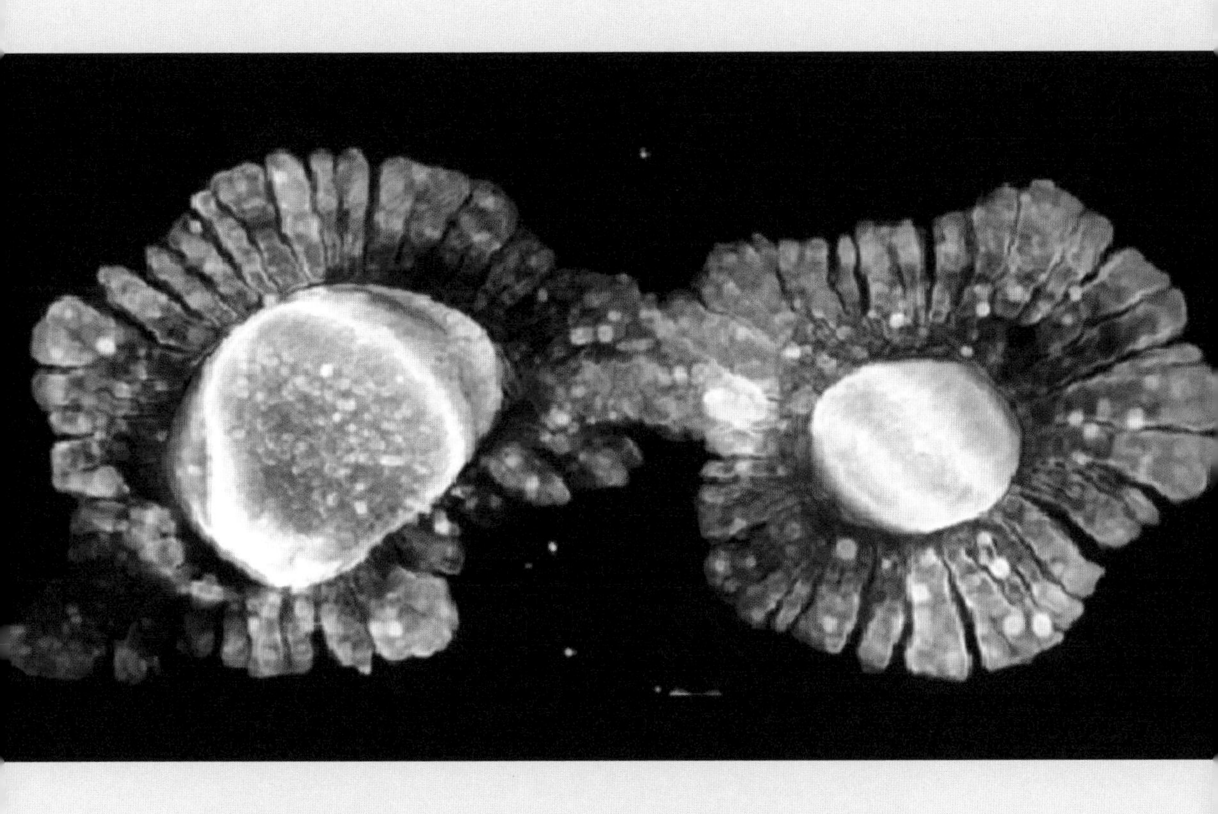

Abderrazak El Albani and others, "The 2.1 Ga [2,100 million year] old Francevillian Biota: Biogenicity, taphonomy and biodiversity," PLOS ONE, vol. 9, 2014; Lobate forms showing sheet-like structure and radial fabric.

**Joanna Liang, Reconstruction of the Cambrian Period arthropod *Habella optata*, 2017.**

The fossils of the Burgess Shale of British Columbia continue to be the subjects of active study, and their affinities to living groups are slowly becoming more apparent. A recent study suggested that this previously enigmatic creature, *Habella*, was probably closely related to the animals that gave rise to spiders and scorpions.

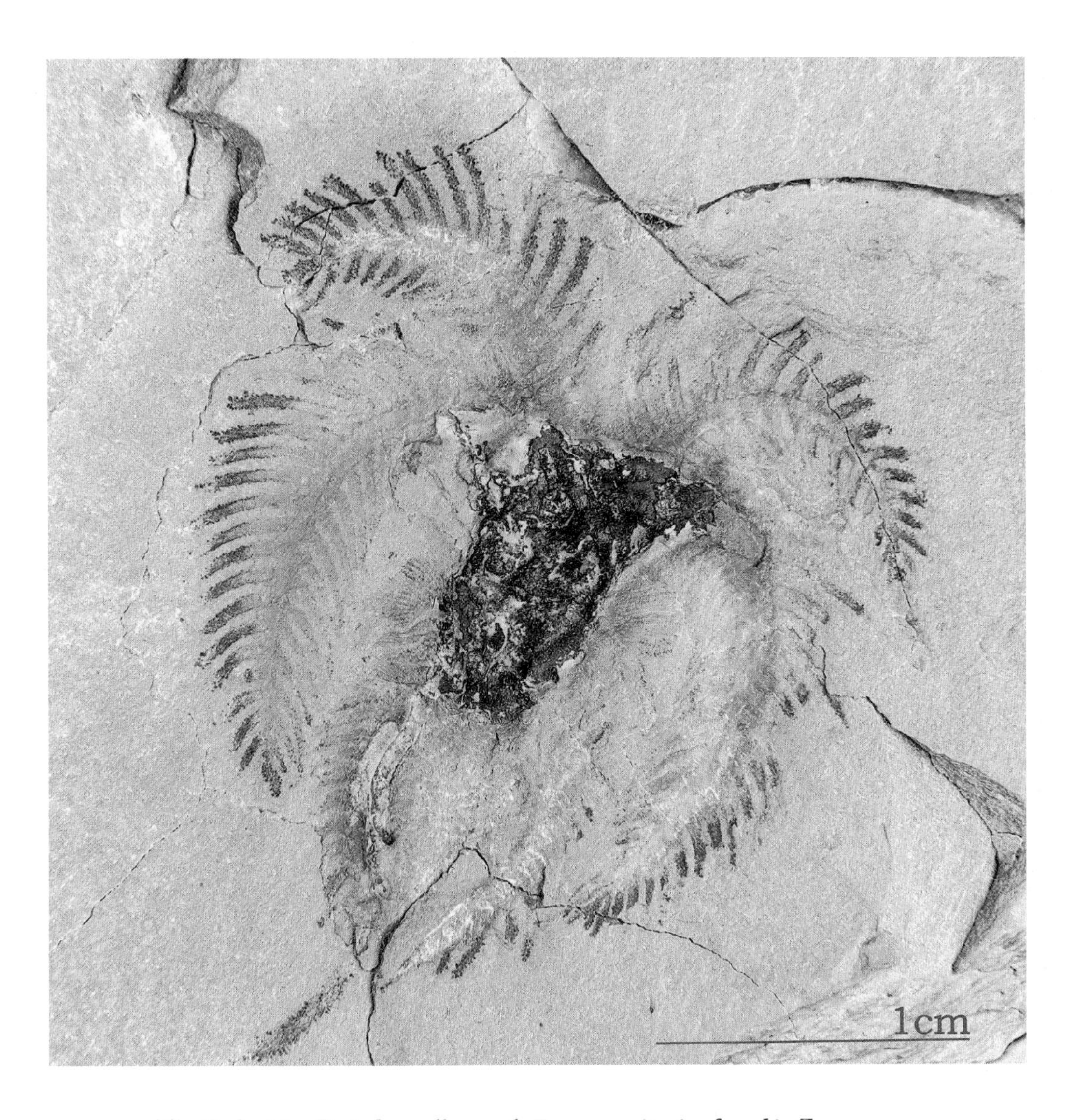


**The Ordovician Period marellomorph *Furca mauritanica*, found in Zagora Province, Morocco, in 2007.**

Halfway across the world from British Columbia, the Fezouata biota are an entirely new assemblage of fossil animals in Morocco, announced in 2010. The Moroccan strata are slightly more recent than the Burgess Shale—they date from 485 to 470 million years ago, and thus "fill in the gap" in the fossil record corresponding to the transition between the diversity of animal forms seen in the Cambrian Period and the establishment in the Ordovician of the major animal groups we see today. Both sets of animals coexist at Fezouata, showing that the older types survived longer and the newer types have more ancient origins than previously thought.

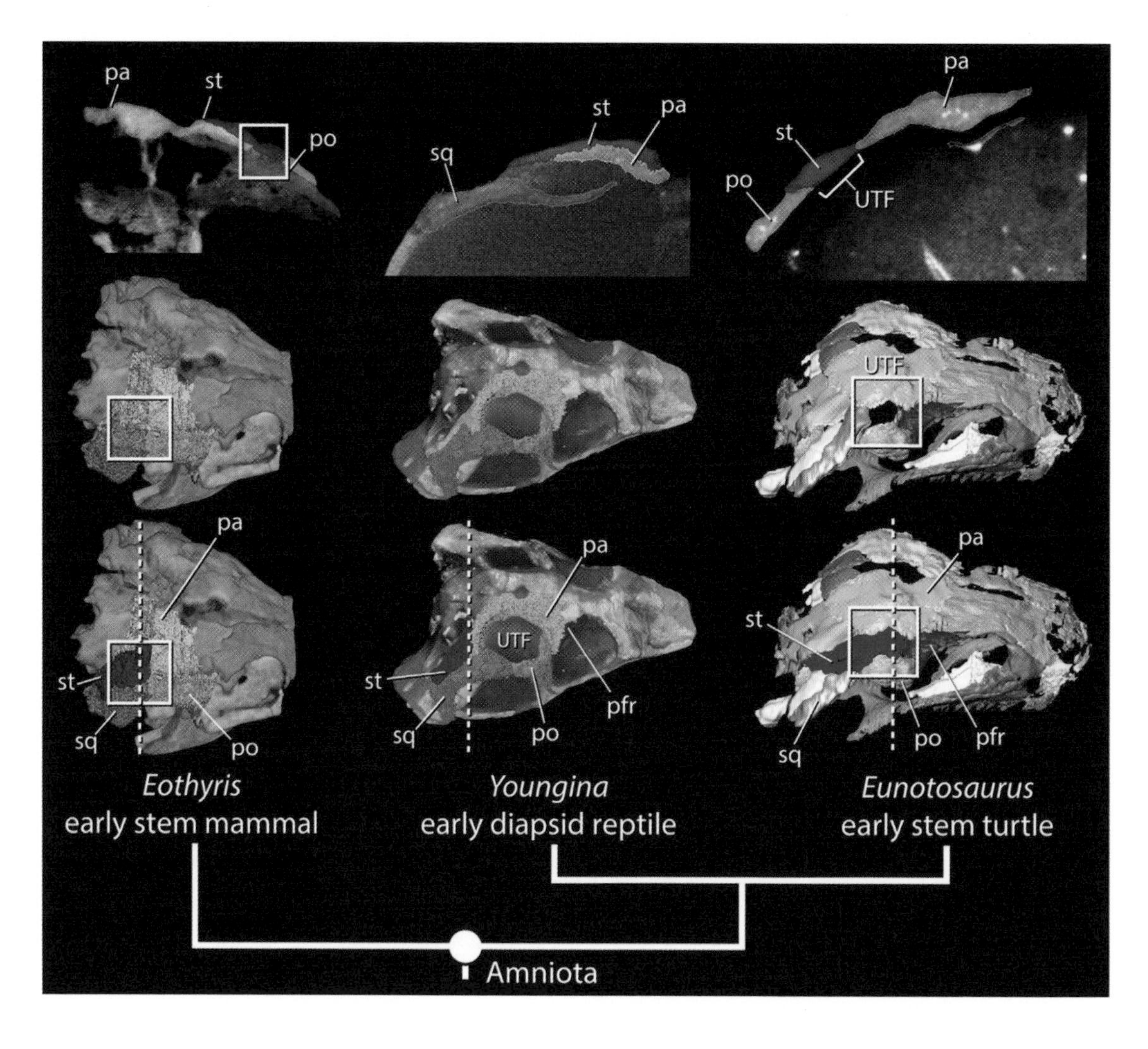


**Gabriel Bever and others, "The amniote temporal roof and the diapsid origin of the turtle skull," *Zoology*, vol. 119, 2016.**

For some time the evolutionary relationships of turtles have been unclear. Turtles do not appear to have muscle-attachment "windows" in the side of their skull, whereas mammals have one, and birds and most reptiles have two. Thus, turtles were assumed to have retained an ancestral arrangement and their links to those other groups remained unclear. This image is from a study which resolves this conundrum by showing that turtles actually evolved from "two-window" forebears—and thus are more closely related to other reptiles than to mammals.

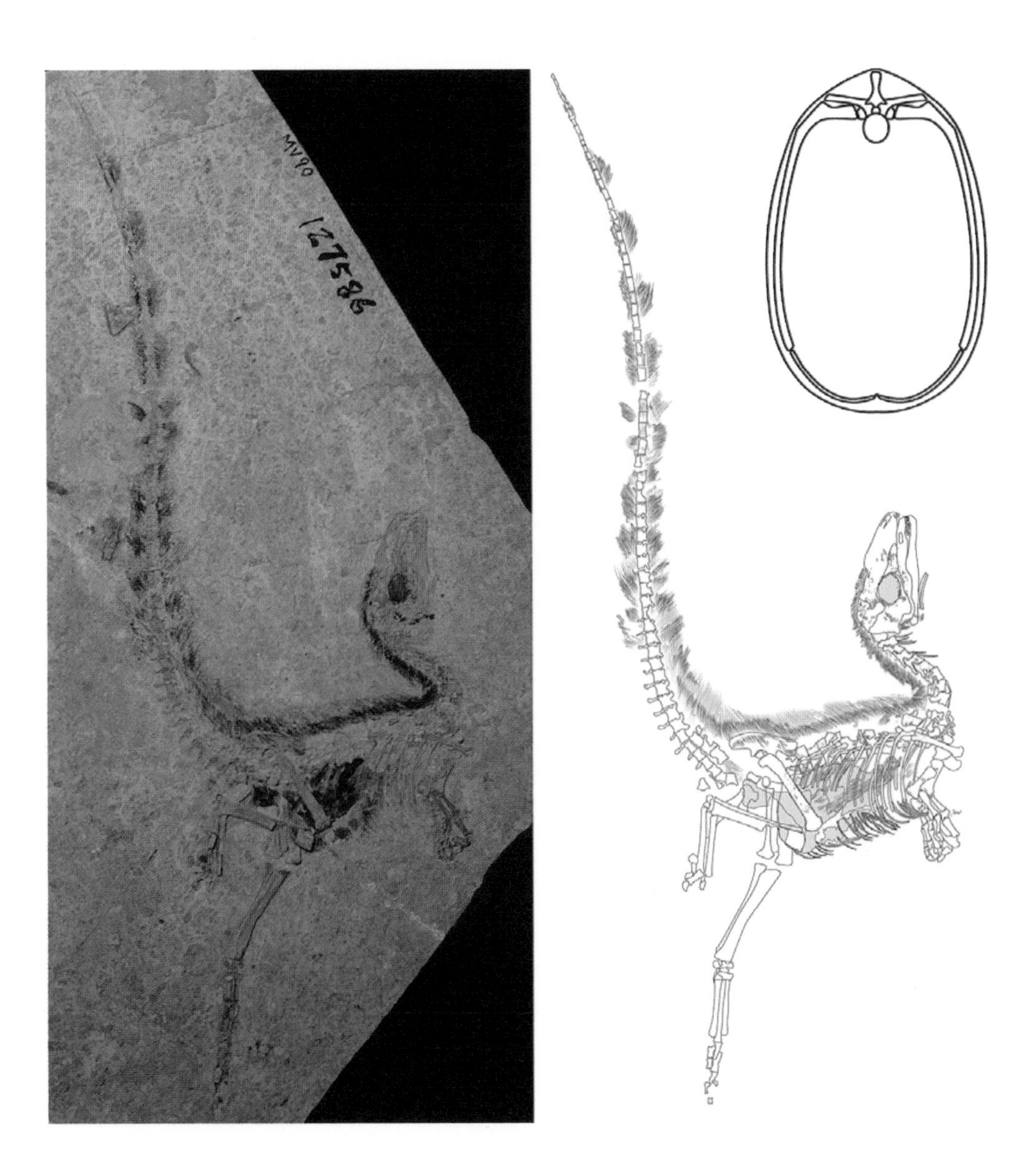


**Fiann Smithwick and others, "Countershading and stripes in the theropod dinosaur *Sinosauropteryx* reveal heterogeneous habitats in the Early Cretaceous Jehol Biota," *Current Biology*, vol. 27, 2017.**

Another striking fossil from China, this dinosaur specimen from the Jehol Formation in the country's northeast retains the black pigment melanin in its feathers. A full reconstruction of its patterning shows that it had a stripy tail, and was countershaded for camouflage—like many animals today its upper surfaces were darker and its lower surfaces were paler. It is estimated to have weighed just over a pound.

Jean Vannier, "Collective behaviour in 480-million-year-old trilobite arthropods from Morocco," Université de Lyon; Linear clusters of the raphiophorid trilobite *Ampyx priscus*.

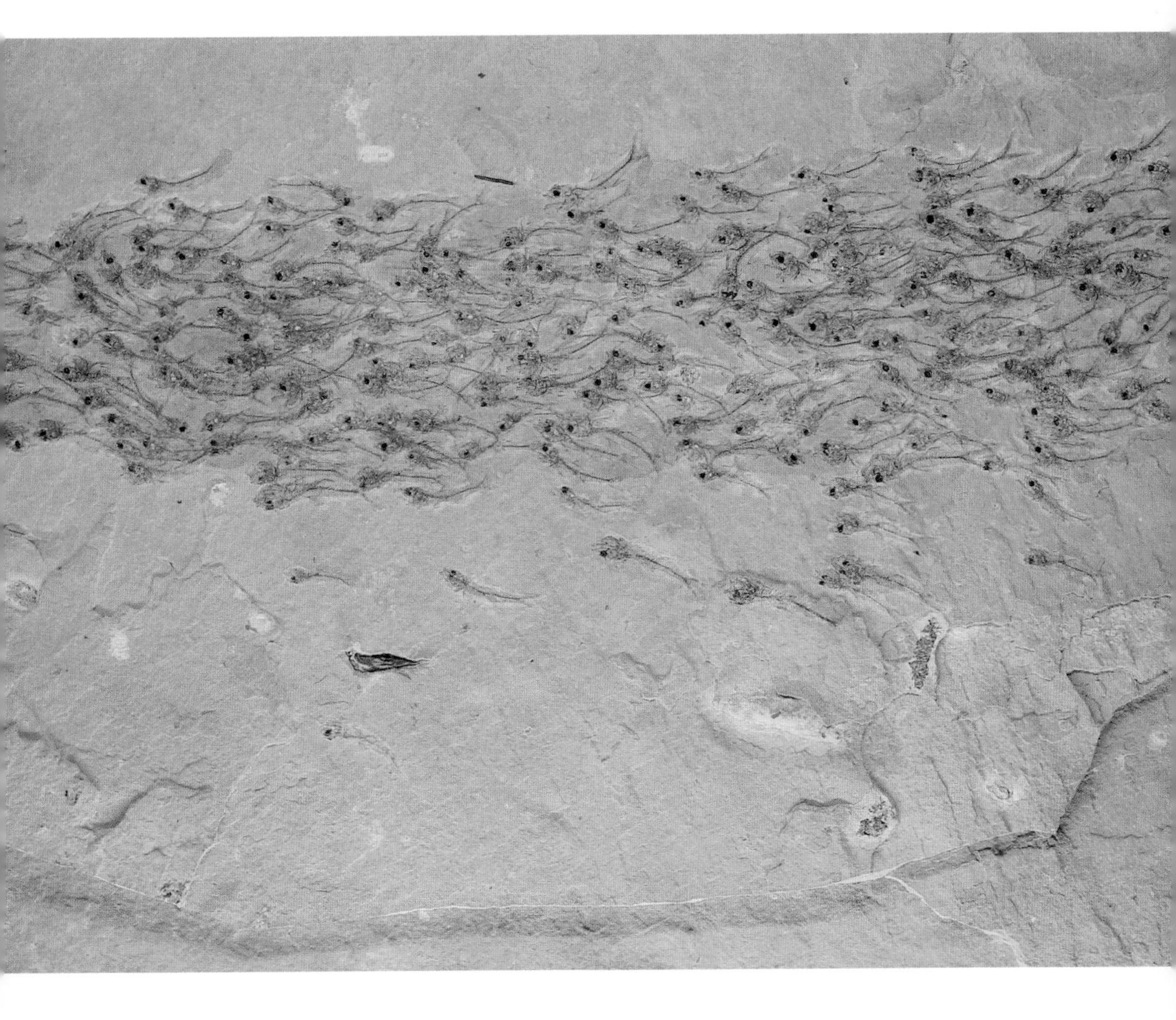

**Proposed fossilized shoal of the Eocene Epoch fish *Erismatopterus levatus*, found at the Green River Formation, Wyoming.**

It had long been assumed that social behavior leaves little trace in the fossil record, but two 2019 studies argue that two very different species—widely separated in time and space—left just such fossils. The opposite image is from a study of 480-million-year-old trilobite fossils aligned in columns, which the authors suggest indicate coordinated migration behavior, or a means of efficient movement through flowing water. The image above is from a mathematical analysis of the orientation and position of the fish in an apparent fossilized "shoal" from the 50-million-year-old Green River Formation of the western United States. In both cases it remains unclear how multiple animals could have died and been preserved in a "snapshot" of communal cooperation.

**Dawid Surmik and others, "Tuberculosis-like respiratory infection in 245-million-year-old marine reptile suggested by bone pathologies," *Royal Society Open Science*, vol. 5, 2018; Histology of the ventral rib of "*Proneusticosaurus*" *silesiacus* holotype in polarized light.**

Animals that fossilize are, of course, dead, and it is becoming increasingly clear that we can determine why some of them died—"paleopathology." The image above is a cross section through a rib of the Triassic Period marine reptile *Proneusticosaurus*, which shows an inflammatory process under the lining of the side of the bone which faces the lungs—a finding typical of tuberculosis, and evidence that this disease has afflicted vertebrates for millions of years. The image opposite demonstrates what a 2019 paper politely described as "evidence of intraspecific agonistic interactions in *Smilodon populator*." The upper canine teeth of saber-tooth tigers are so unwieldy that many have argued they could only be used for display, but two *Smilodon* individuals have been discovered with holes in their skulls which match the size, shape, and other likely features of penetration by another *Smilodon*'s fang.

**Opposite: Skulls of two Pleistocene Epoch saber-tooth tigers, *Smilodon populator*, mounted to show bite injury.**

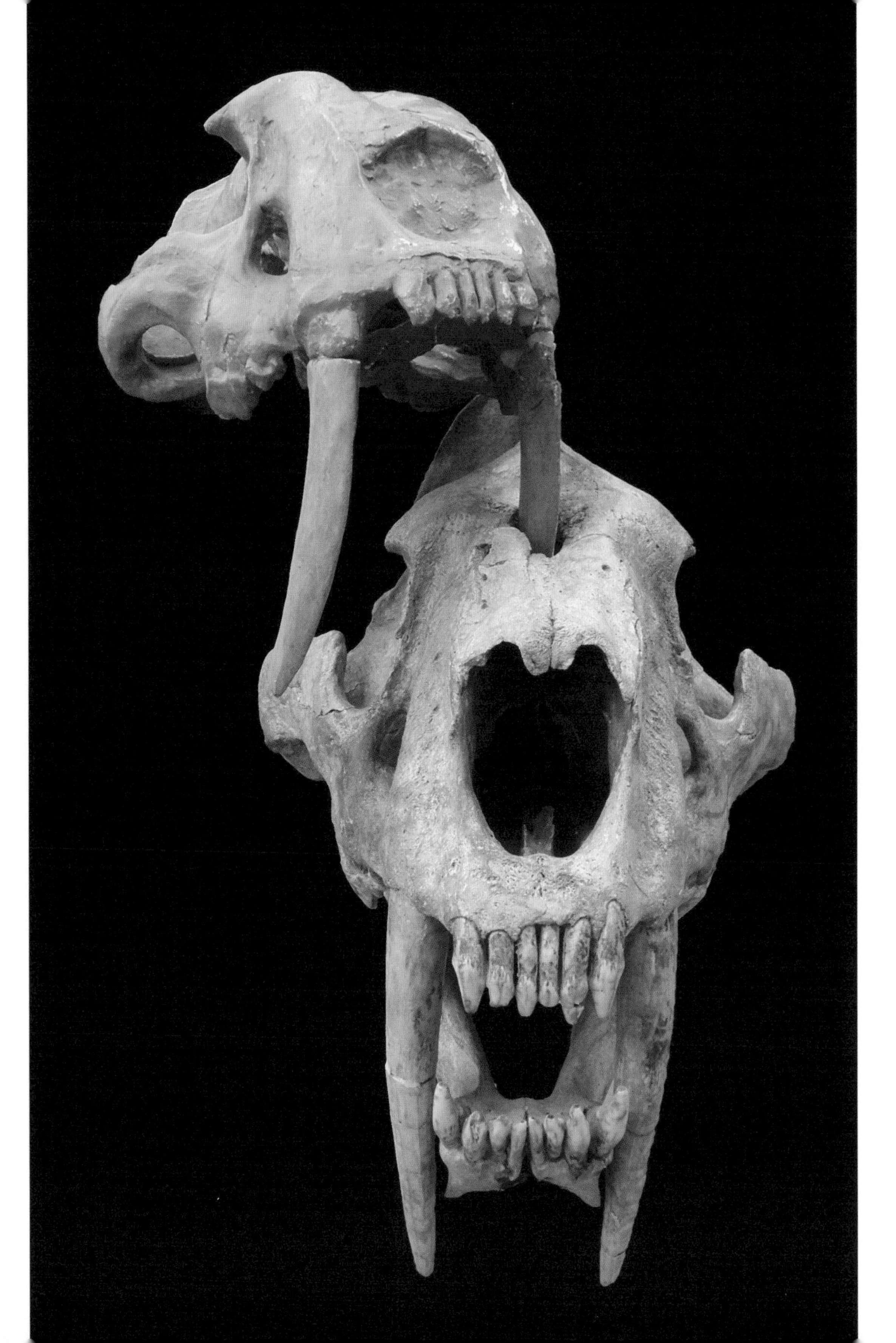

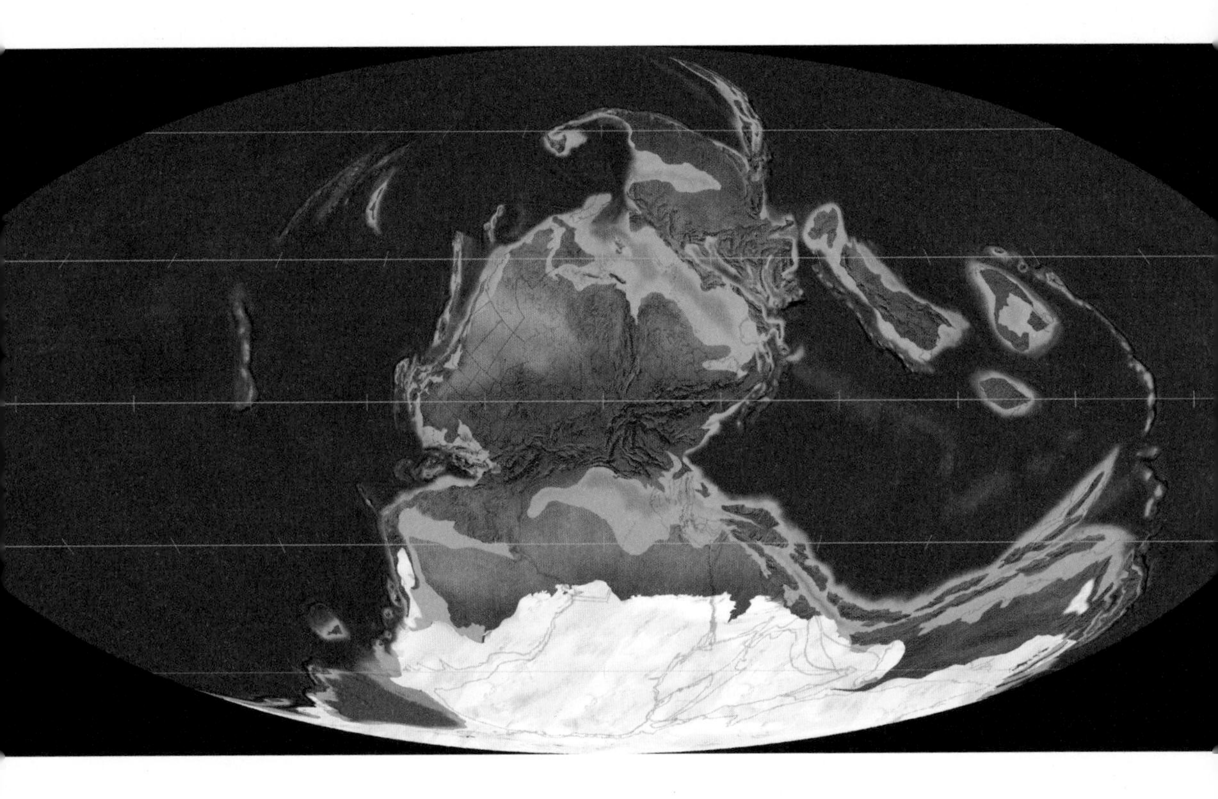

**World map at time of Carboniferous Rainforest Collapse, 300 Ma, late Pennsylvanian era. Map by Ron Blakey, Colorado Plateau Geosystems Inc.**

The Carboniferous Rainforest Collapse was a period around 307 million years ago when the world's climate changed dramatically, leading to pronounced environmental, atmospheric, and ecological upheaval. In general the climate became drier, leading to the loss of the rainforests spanning the equatorial supercontinent that comprised modern Eurasia and North America. These were the famous "coal forests" of the Carboniferous Period, which indirectly increased atmospheric oxygen concentrations to their highest-ever levels—35 percent compared to today's 21 percent—allowing insects, for example, to grow to unprecedented size. Thus, the Carboniferous Rainforest Collapse was an extinction event, but one which created the conditions conducive for future successes already waiting in the wings: flowering plants, reptiles, and mammals.

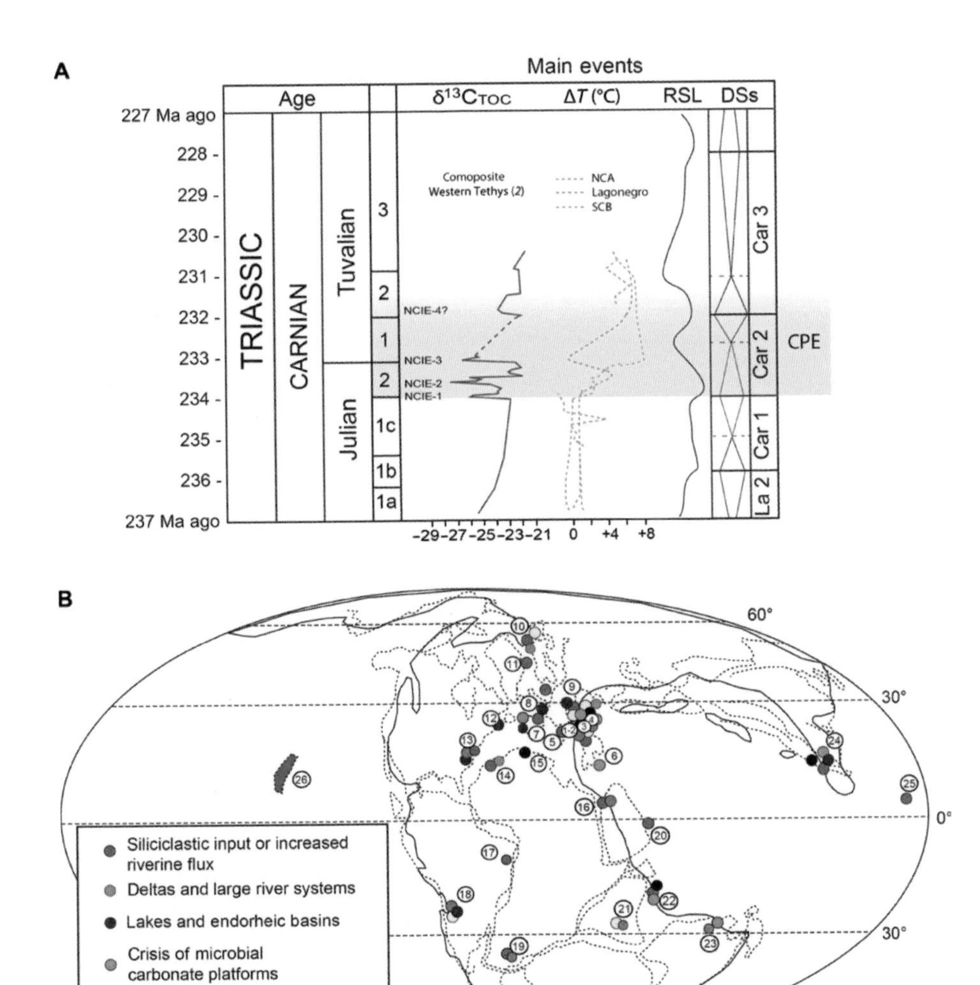


## Jacopo Dal Corso and others, "Extinction and dawn of the modern world in the Carnian (Late Triassic)," *Science Advances*, vol. 6, 2020.

The Carnian Pluvial Episode, 233 million years ago, was a later extinction event which occurred when the world became wetter. There were four global episodes of increased rainfall, possibly resulting from global warming, in turn possibly caused by an epoch of spectacular volcanism in what is now western Canada. Although the climate changed from dry to wet—the opposite of the Carboniferous Rainforest Collapse—the general effect on life was somewhat similar. There was a loss of many groups of plants and animals, but a subsequent diversification of several pre-existing groups: conifers, insects, mammals, and most reptiles, including dinosaurs.

**Above: Reconstruction of *Juravenator starki* by Tom Parker, 2015.**

*Juravenator* was an early dinosaur which lived in Bavaria 151 million years ago, and is known only from one juvenile specimen just over 2ft (60cm) in length. The species may have been covered with feathers over most of its body, but most notable is a band of scales running along the top of its tail. These scales each bear a tiny dome almost identical to those found on the sensory scales of modern crocodiles and alligators, suggesting that *Juravenator*'s tail was a featherless sensory organ. Other evidence hints that this species was nocturnal, so it has been speculated that it dipped its tail in bodies of water to detect the wriggling of potential prey in the dark of night.

**Left: Holotype fossil of juvenile *Juravenator starki*, found at Kieselplattenkalken Schamhauptenm Solnhofen archipelago, Late Jurassic era. Jura-Museum Eichstätt.**

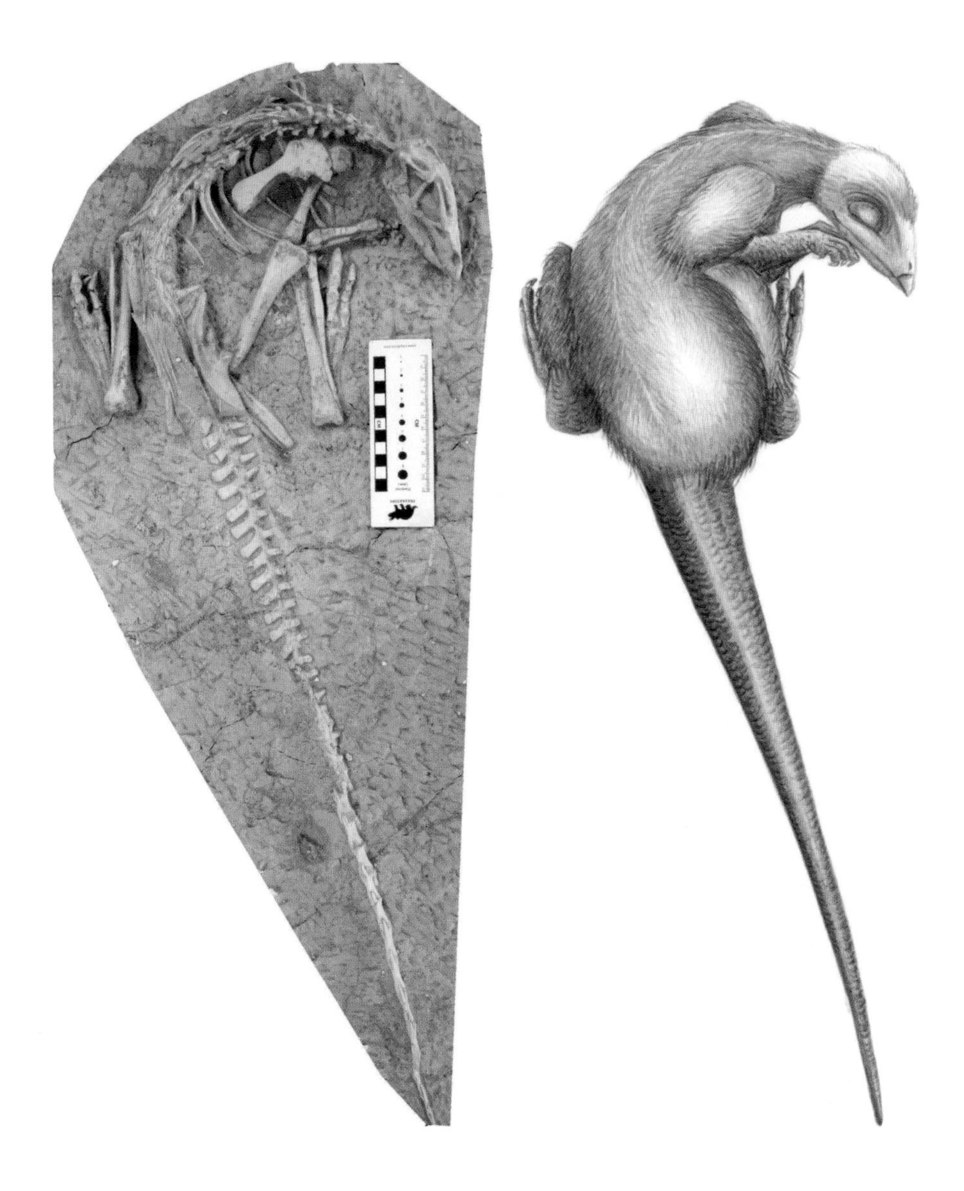

**Yuqing Yang and others, "A new basal ornithopod dinosaur from the Lower Cretaceous of China," *PeerJ*, vol. 8, 2020; Holotype of *Changmiania liaoningensis* in dorsal view (left) and reconstruction by Carine Ciselet/ RBINS-IRSNB-KBIN (right).**

Another diminutive dinosaur fossil, this time from Liaoning Province in China and dating from 125 million years ago, provides an insight into an ancient way of life. A complete skeleton of *Changmiania liaoningensis* was found in a relaxed "sleeping position," with no sign of scavenging or erosion of its remains. Its anatomy has been interpreted to indicate that this dinosaur dug and lived in burrows—it possessed a robust snout, short neck, and the large, knobbly shoulder blades typical of burrowing species, and its back end shares some features with moles. Thus, the authors of this paper suggest that this creature was killed instantaneously when a volcanic eruption caused its burrow to collapse.

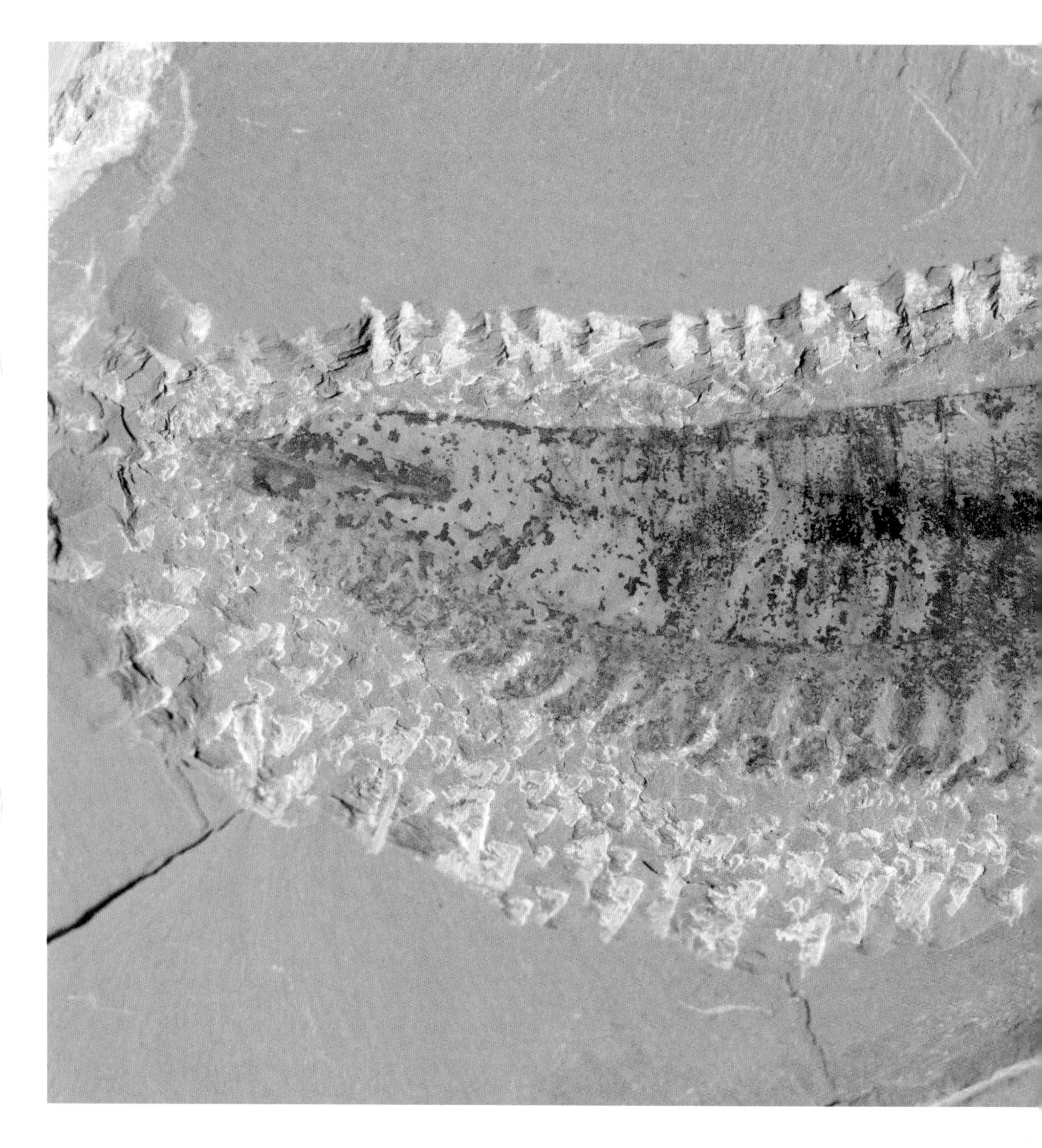

**Han Zeng and others, "An early Cambrian euarthropod with radiodont-like raptorial appendages," *Nature*, vol. 588, 2020; Anatomy of *Kylinxia zhangi* found in the early Cambrian Chengjiang biota, in Yunnan Province, China.**

Arthropods—insects, spiders and scorpions, millipedes and centipedes, and crustaceans—are one of the most diverse and abundant groups of animals, yet their origins are somewhat tangled. The Burgess Shales (see page 118) yielded several fossils now

thought to be related to modern arthropods, but their array of patterns of segmentation, appendages, and mouthparts makes it difficult to unravel their interrelationships. This species, *Kylinxia zhangi* from the early Cambrian Chengjiang biota (see page 226), possesses a combination of features which has helped to establish structural equivalences between fossil and living forms—in other words, which body parts in one arthropod species correspond to which body parts in another.

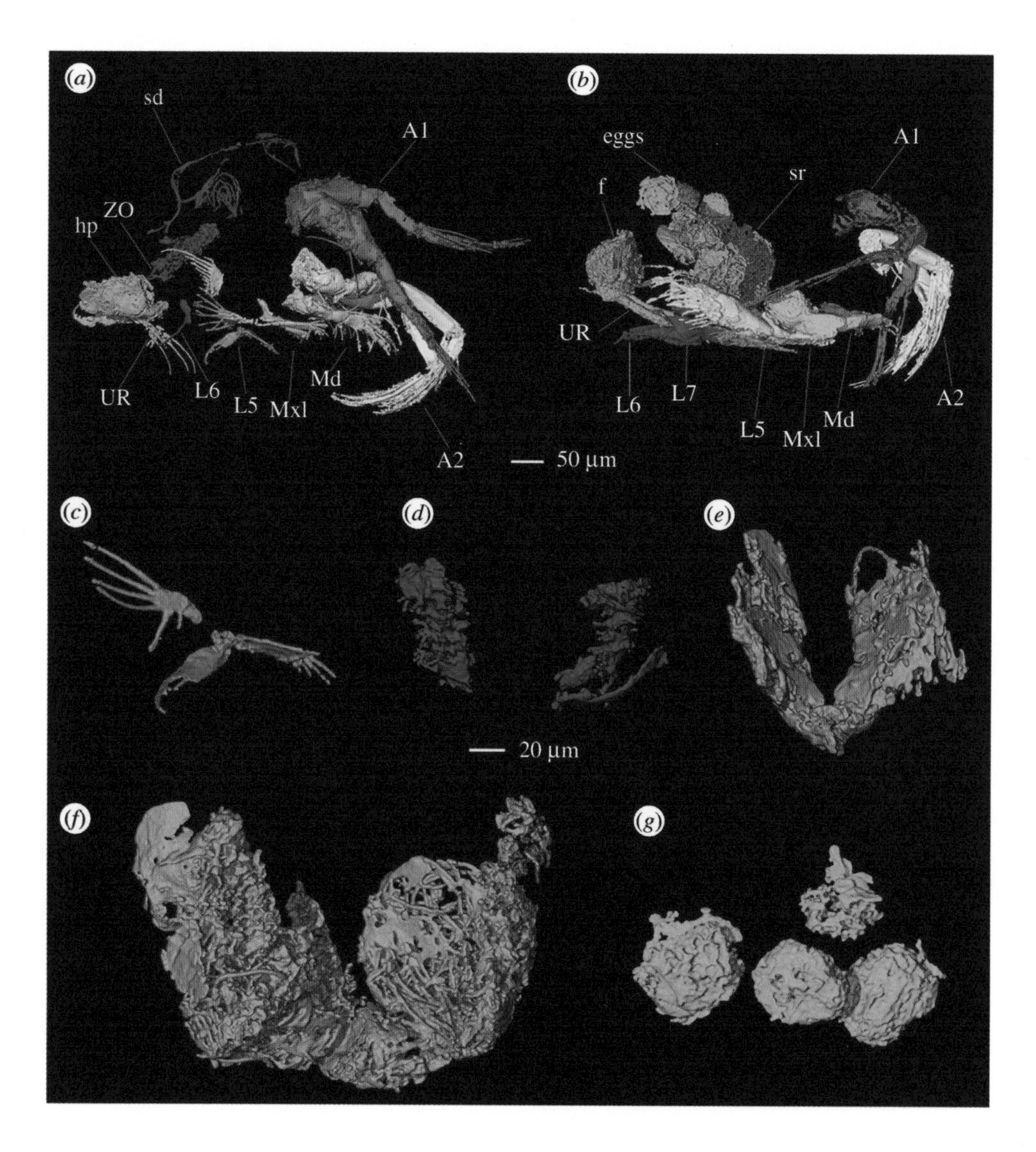


**He Wang and others, "Exceptional preservation of reproductive organs and giant sperm in Cretaceous ostracods,"** ***Philosophical Transactions of the Royal Society of London*****, Series B, vol. 287, 2020; Soft parts of** ***Myanmarcypris hui*****.**

Ostracods are unusual crustaceans which live inside a two-piece "shell" made of the same material as crustacean and insect cuticles, chitin. They have a long evolutionary history dating back to the original radiation of arthropods in the Ordovician Period. The 100-million-year-old specimens depicted here have been scanned by computed tomography to reveal the three-dimensional shape of their internal organs: a male at upper left and a female at upper right. The different tissues are then color-coded and "virtually dissected"—for example, the tangled mauve structure at lower left is the female's seminal receptacle, full of giant sperm. The breeding habits of ostracods have not changed much since.

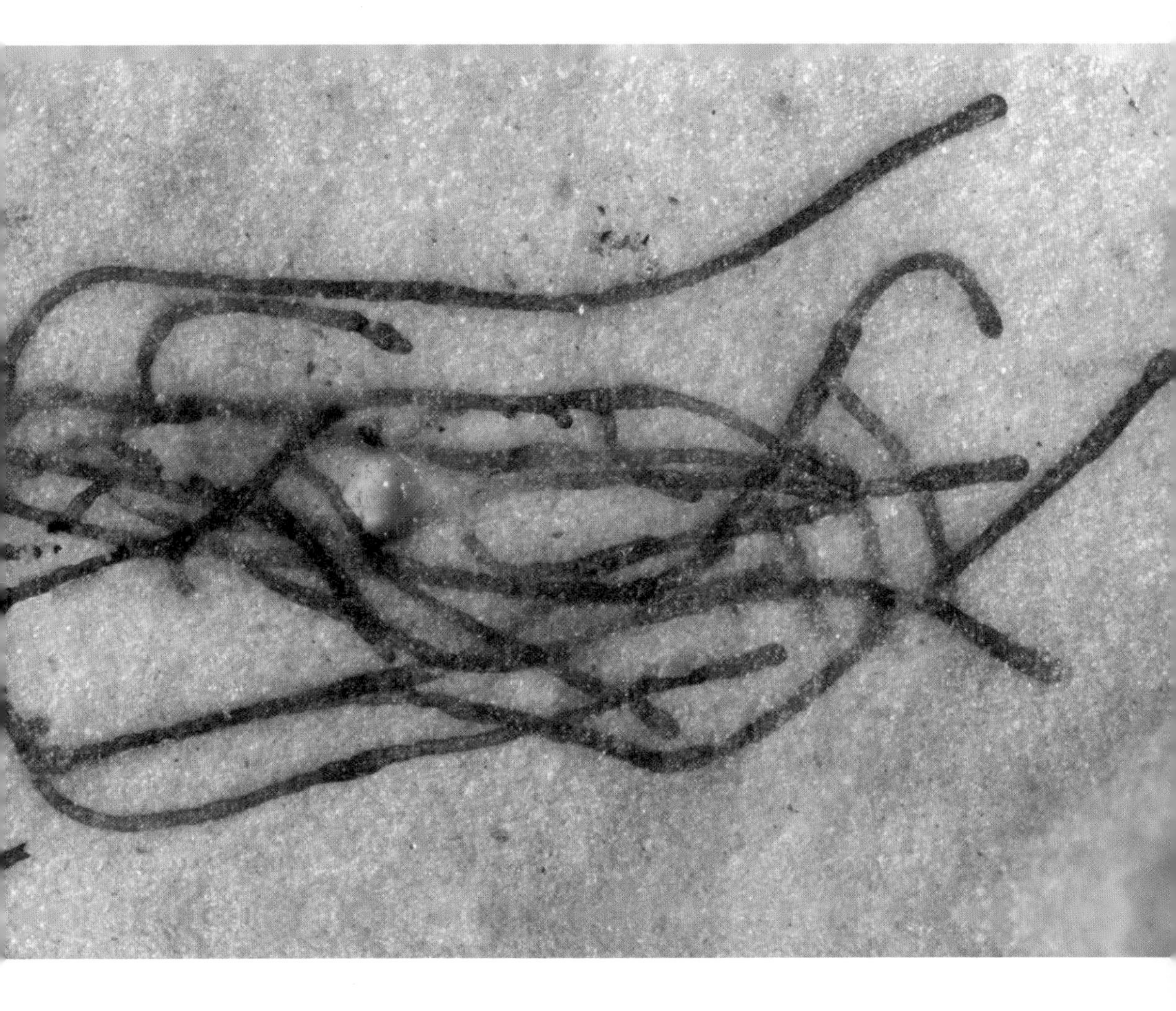

**Qing Tang and others, "A one-billion-year-old multicellular chlorophyte," *Nature Ecology & Evolution*, vol. 4, 2020; Cellular structures of the algae *Proterocladus antiquus*, a new species.**

A key point in evolutionary history came when a photosynthesizing cyanobacterium was incorporated into a nucleated host cell, thus creating the first plant. This specimen is a sprig of algae from 1,000 million years ago, around the time when the increase in atmospheric oxygen concentrations accelerated to attain levels similar to today's.

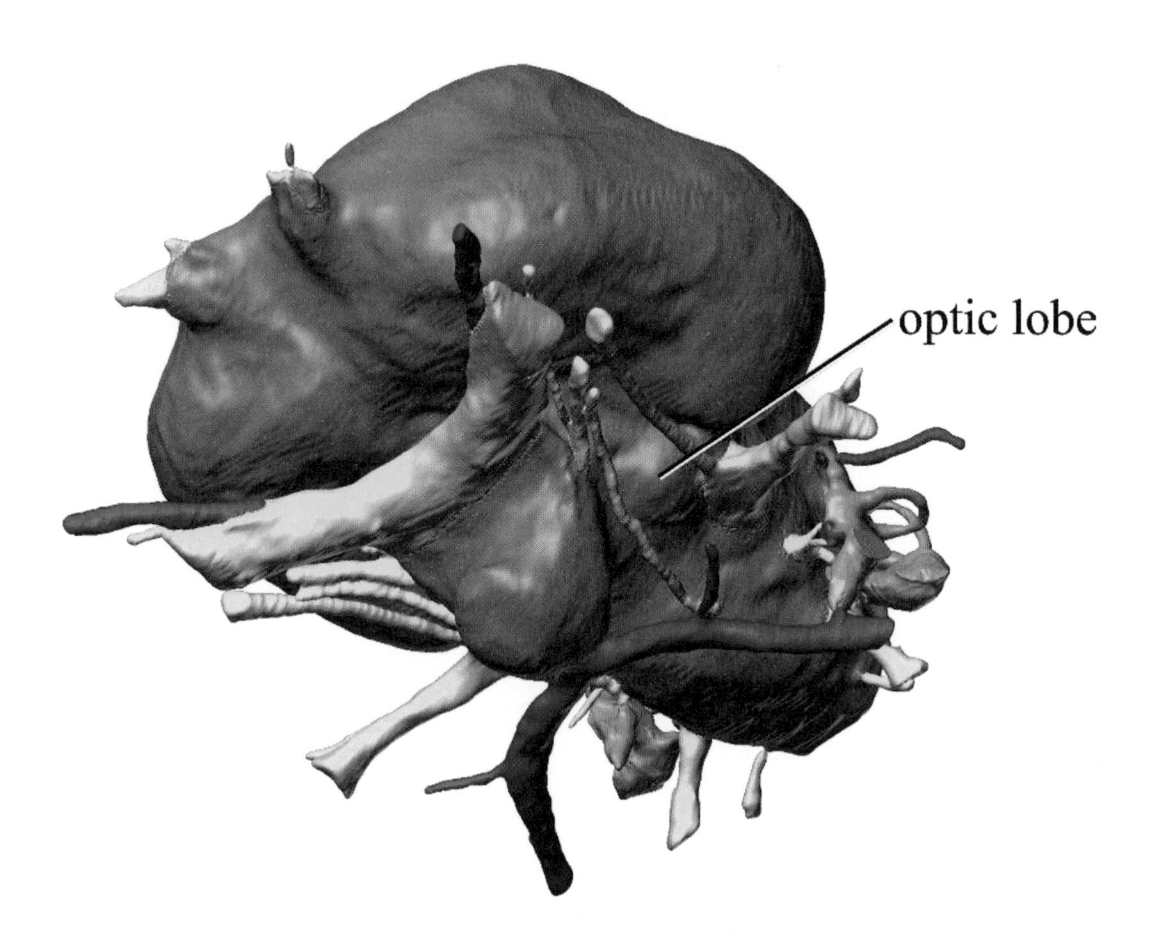


**Catherine Early and others, "Beyond endocasts: using predicted brain-structure volumes of extinct birds to assess neuroanatomical and behavioral inferences,"** ***Diversity*, vol. 12, 2020; Endocast of *Dinornis robustus*.**

There is a long history in paleontology of making casts—"endocasts"—of the insides of fossil skulls to estimate the size and shape of the brain. Studies of modern birds have shown that the internal shape of the skull indeed parallels the size and shape of the brain regions it encases, and this means that computed tomography can now be applied to fossil skulls to measure those same brain regions. This image is a computer-generated "cast" of the brain of an extinct moa (see page 105). This moa, for example, had unusually small optic lobes—so perhaps a giant flightless bird with no predators does not need to see as well as other birds.

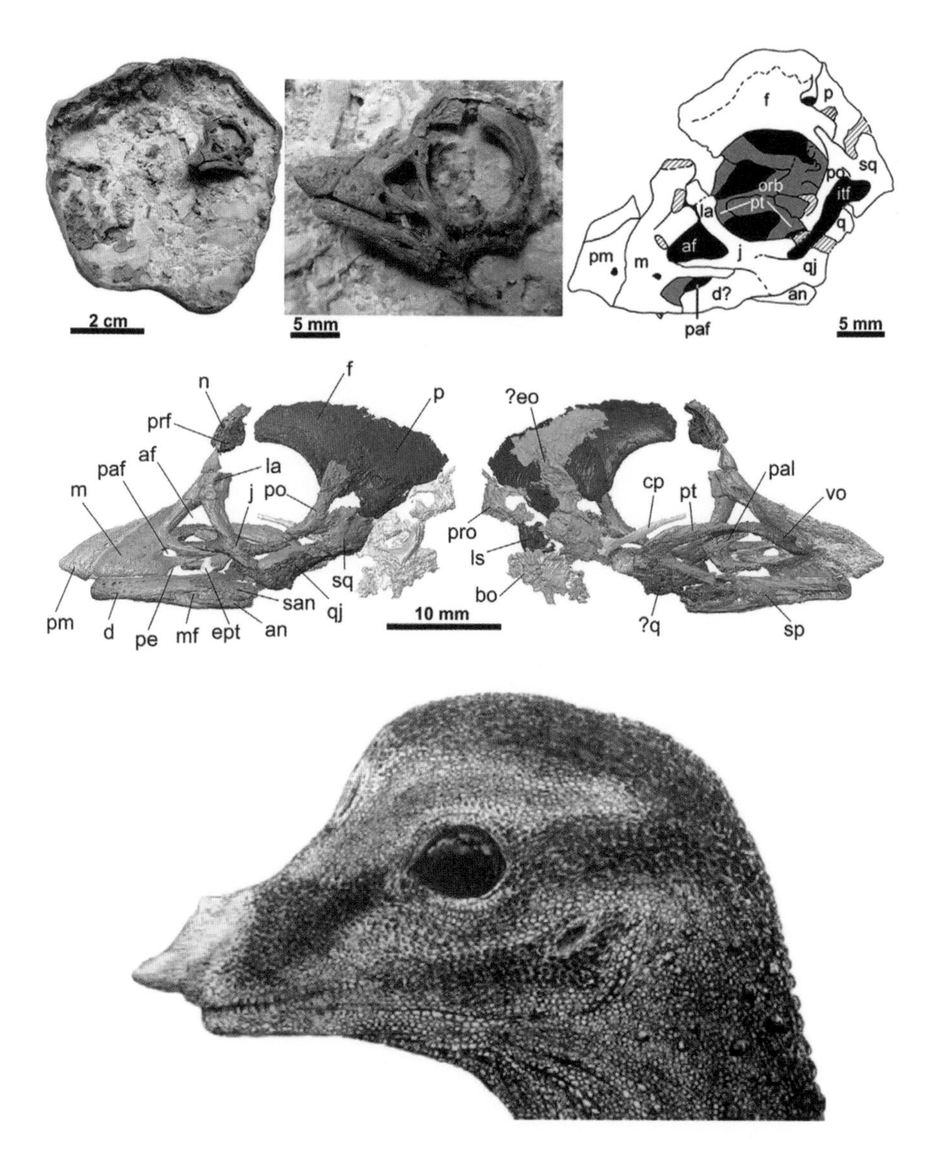


**Martin Kundràt, "Specialized craniofacial anatomy of a Titanosaurian embryo from Argentina," *Current Biology*, vol. 30, 2020.**

Computed tomography allows paleontologists to study phenomena which would have seemed unimaginable just a few decades ago. These images are from a CT study of an embryo of a giant sauropod (in other words "brontosaurus-shaped") while still in its egg. This "paleoembryology" shows that baby titanosaurs had unusual tiny horns on their snouts, connected to a series of bony channels perhaps containing sensory organs. These infants' eyes were angled more toward the front than most dinosaurs, suggesting they possessed stereoscopic vision—an adaptation that allows animals, including cats and primates, to accurately calculate the distance of objects such as food items or predators.

# CREDITS

*Every attempt has been made to trace the copyright holders of the works reproduced, and the publishers regret any unwitting oversights. Illustrations on the following pages are generously provided courtesy of their owners, their licensors, or the holding institutions as below:*

Cover images: Wellcome Collection, London (CC BY 4.0) top; Smithsonian Libraries, Washington, courtesy of (via BHL) bottom

A. Albani et al, "The 2.1 Ga Old Francevillian Biota," *PLOS ONE*, 2014, 9(6),fig. 3a–b: e99438. https://doi.org/10.1371/journal.pone.0099438: 231

*Abhandlungen der Königlichen Bayerischen Akademie der Wissenschaften*, 1936, vol. 33, p.65, fig. 8, Abhandlungen:N.F. 33, after, http://publikationen.badw.de/en/006296956 (CC by-4.0): 151

Academy of Natural Sciences of Drexel University, Philadelphia, ANSP Archives Collection 803: 64 top

Alamy Stock Photo: 40 (Album / The British Library); 87 (Azoor Photo); 158 (Dotted Zebra); 113 top (The History Collection); 109 (Historic Images); 25 (Jimlop collection); 80 (Quagga Media); 113 both (Museum für Naturkunde, Berlin/History and Art Collection); title page, 49, 164–165 (Natural History Museum, London/ Science History Images); 196–197 bottom (Nicolas Fernandez); 154 (Universal Images Group North America LLC / De Agostini Picture Library); 204 (Universal Pictures/Allstar Picture Library Ltd.)

American Museum of Natural History Library, New York: 108, 114

American Museum of Natural History, New York/photo: Smokeybjb, via Wikimedia Commons (CC by-SA 3.0): 136

Amgueddfa Genedlaethol Cymru / National Museum of Wales: 42–43

© 2018 The Authors, Royal Society Publishing (CC BY 4.0): 238

Beijing, Chinese Academy of Sciences (rights holder unknown): 200

Biodiversity Heritage Library: 97

Blok Graphic, diagram after J. N. Keating et al, with illustrations via Wikimedia Commons (except where stated)—*Cyclostomi*: NOOAA Great Lakes Environmental Research Laboratory, via Flickr (CC BY-SA 2.0); *Ar, andaspida, Anaspida* , and *Galeaspida*: Nobu Tamura (CC by-4.0) http://spinops.blogspot.com/2018/09/cowielepis-ritchiei.html; *Heterostraci* and *Thelodonti*: Matteo De Stefano/MUSE (CC BY-SA 3.0); *Astrapsis*: Philippe Janvier, 1997 http://tolweb.org/Astraspida/16906; *Osteostraci:* Ghedo (CC BY-SA 4.0); *Antiarchi*: Citron (CC-BY-SA-3.0); *Arthrodira*: Dmitry Bogdanov (CC by 3.0); *Chondrichthyes*, James St John, via Flickr (CC by 2.0); *Sarcopterygii*, Alberto Fernandez Fernandez (CC BY 3.0); *Actinopterygii*: Diego Delso (CC BY-SA 4.0): 220–221

© The British Library / Bridgeman Images: 53

The British Library (CC0 1.0): 24

California Academy of Sciences Library (via BHL): 20–21, 23, 57

Cambridge University Press, after Dennis M. Bramble, "Origin of the mammalian feeding complex," *Paleobiology*, vol. 4, issue 3, Summer 1978, fig. 6 (detail), reproduced with permission: 180

Carine Ciselet/RBINS-IRSNB-KBIN: 243

Carnegie Museum of Natural History, Copyright ©: 99

Carol Hotton, Dr, courtesy of, after F. Hueber et al, "Devonian Landscape Heterogeneity Recorded by a Giant Fungus," *Geology*, 2007, no. 35, issue 10.1130: 213

Cleveland Museum of Natural History, photo: Tim Evanson, Cleveland Heights, Ohio, via Flickr (CC BY-SA 2.0): 105

Copyright © 2020 The Authors, American Association for the Advancement of Science, Open access (CC BY 4.0), from J. Dal Cordo et al, "Extinction and dawn of the modern world in the Carnian (Late Triassic)," *Science Advances*, 16/09/2020, vol. 6, no. 38, fig. 1: 241

Colin Palmer, courtesy of (illustration James Brown): 253

*Current Biology* 27, 3337–3343, Nov. 6, 2017 © 2017 The Authors. Elsevier Ltd. https://doi.org/10.1016/j.cub.2017.09.032: 235

David Bainbridge, University of Cambridge: 18, 44, 45, 46, 110, 128, 131, 132, 136, 137, 138, 139, 159

Didier Descouens, via Wikimedia Commons (CC BY-SA 3.0): 233

Dirk Fuchs, Dr, courtesy of, after D. Fuchs et al, "New Octopods (Cephalopoda: Coleoidea) From The Late Cretaceous (Upper Cenomanian) Of Hâkel And Hâdjoula, Lebanon," *Palaeontology*, 2008, vol. 52, issue 1, fig.2: 222

Dominic Papineau and Matthew Dodd, Drs, courtesy of: 228

D. M. Raup and J. J. Seposki, "Mass Extinctions in the Marine Fossil Record," *Science*, 1982, vol. 215, issue 4539, page 1501, fig. 1. The American Association for the Advancement of Science. Reprinted with permission from AAAS: 192

D-Y Huang and H. Zeng, Nanjing Institute Of Geology And Palaeontology, Singapore: 244–245

Eduard Solà, 2012, via Wikimedia Commons (CC BY-SA 3.0): 212

Eduardo Ascarrunz; Jean-Claude Rage; Pierre Legreneur; Michel Laurin, courtesy of, via Wikimedia Commons (CC BY 3.0): 149

Erlend Bjørtvedt, via Wikimedia Commons (CC BY-SA 4.0): 208

*Fossil Record*, vol.20, issue 10.5194/fr-20-129-2017, fig.7, (CC by 3.0): 167

Francis Latreille, courtesy of: 186

Fukui Prefectural Dinosaur Museum, Katsuyama, Japan, Courtesy: 237

Gabriel S. Bever, Center for Functional Anatomy and Evolution, John Hopkins School of Medicine, courtesy of: 234

Getty Images: 176 (Alain Nogues/Sygma); 16, 115, 129, 141 (Bettmann Archive); 205 (Ernest Bachrach/John Kobal Foundation); 65 (Fox Photos); 202 (John Weinstein/Field Museum)

Getty Research Institute, Los Angeles: 167 bottom

Ghedoghedo, 2010, via Wikimedia Commons (CC BY-SA 3.0): 193

Gozitano, 2016, via Wikimedia Commons (CC BY 4.0): 62

Günter Bechly, Dr, via Wikimedia Commons (CC BY-SA 3.0): 189

Harvard University, Museum of Comparative Zoology, Ernst Mayr Library (via BHL): 1, 48, 51, 59, 61, 86 right, 90, 107, 122–123, 172–173

Houston Museum of Natural Science, Courtesy of the; 2012.1933.01/Cat No. PB.440: 229

H. Zell, 2010, via Wikimedia Commons (CC BY-SA 3.0): 108

Illustration by Dr He Wang, by kind permission: 246

Illustrations from *The Giant Golden Book of Dinosaurs and Other Prehistoric Reptiles* by Jane Werner Watson, copyright © 1960 by Penguin Random House LLC. Used by permission of Golden Books, an imprint of Random House Children's Books, a division of Penguin Random House LLC. All rights reserved.: 161

Isaac Sanchez, via Flickr (CC 2.0): 133

Iziko Museum of Natural History (University of the Witwatersrand): 73

Ji í Burian, for the estate of Zdenek Burian: 160

J. Vannier (Univ. Lyon 1): 236

J-M. R. Maurrasse, Courtesy of; after J-M. R. Maurrasse et al, "Spatial and Temporal variations of the Haitian K/T Boundary record," *Journal of Iberian Geology*, 2004, vol. 31, pp. 113–133, fig. 4, with thanks to Frank Asaro: 179

© James St. John, via Flickr (CC BY 2.0): 102

Jenny Clack, Professor; courtesy of Rob Clack: 184, 194, 196 top, 197 top

Jonathan Chen, via Wikimedia Commons (CC BY-SA 4.0): 201

Jose Manuel Canete, via Wikimedia Commons CC BY-SA 4.0): 199 bottom

Junnn11, via Wikimedia Commons (CC by-SA 4.0): 226 bottom

Leicester Museums & Galleries: 156–157

Library of Congress, Washington D.C., Prints and Photographs Division: 142

Linda Hall Library of Science, Engineering & Technology, courtesy of the (CC BY 4.0): 96

Lunar and Planetary Institute, Courtesy of David Kring: 178

Luxquine, via Wikimedia Commons (CC BY-SA 4.0): 12

© Maja Daniels: 195

Martin Lipman © Canadian Museum of Nature: 218–219

Maryland Center for History and Culture, Baltimore, courtesy of the: 47 (Item RS3905)

Matteo De Stefano/MUSE, via Wikimedia Commons (CC by 3.0): 170 top

MFA, Boston: Photograph © 2021 Museum of Fine Arts, Boston, Helen and Alice Colburn Fund, 63.420: 6

M. P. Witton and M. B. Habib, from "On the Size and Flight Diversity of Giant Pterosaurs," *PLOS ONE* 5(11): e13982, 2010, fig. 6. 5(11): e13982; https://doi.org/10.1371/journal.pone.0013982: 188

Museo Archeologico di Altamura: Soprintendenza Archeologia della Puglia: 203

© Museum für Naturkunde, Berlin: 86 left (H. Raab, via Wikimedia Commons (CC BY-SA 3.0)), 76 (VeryFullHouse (GNU General Public License))

Natasha de Vere & Col Ford, via Flickr (CC by 2.0): 63 bottom

National Central Library of Florence (PDM 1.0): 85

*National Science Review*, Jim Meng, "Mesozoic mammals of China," 2014, vol. 1, p.534, fig. 7 (CC by 4.0): 181

Natural History Museum, Bamberg/Photo: Reinhold Möller, 2017: 72

Natural History Museum Library, London (via BHL): 30, 71, 77, 81, 89, 104, 130

Naturalis Biodiversity Center, Leiden: 50

NRF-SAIAB National Fish Collection. Photo: Chip Clark, Courtesy of the Smithsonian Institution: 72 bottom

Nicolás R. Chimento, courtesy of: 239

Open access © 2020 by the authors (CC By 4.0), from *Diversity*, 2020, 12(1), issue 34, fig. 7; https://doi.org/10.3390/d12010034: 252

Opensource text, via archive.org (CC BY-SA 3.0): 125

© Oxford University Museum of Natural History: 35, 46, 52 bottom

Peabody Museum of Natural History, Yale University, New Haven, Connecticut, USA, courtesy of the: 98, 100; 191; *The Age of Mammals*, a mural (detail) by Rudolph F. Zallinger. Copyright © 1966, 1975, 1989, 1991, 2000 Peabody Museum of Natural History, Yale University, New Haven, Connecticut, USA: 169

Pennsylvania Academy of Fine Arts; Gift of Mrs Sarah Harrison (The Joseph Harrison, Jr. Collection): 19

PNAS, from A.R.Schmidt et al, "Arthropods in amber from the Triassic Period," *PNAS*, September 11, 2012, vol. 109 (37), fig. 3 (detail): 227

Private collection: 64 bottom

Qing Tang and Shuhai Xiao, Virginia Tech College of Science, 2020: 251

Reprinted from M. Kundràt et al, "Specialized Craniofacial Anatomy of a Titanosaurian Embryo from Argentina," *Current Biology*, November 2, 2020, vol. 30, issue 21, figs 1a–e, 4b, with permission from Elsevier: 253

Reprinted from *Sciencedirect*, vol. 580, issue 1, M. Dierick et al, "Micro-CT of fossils preserved in amber," p.643, fig. 2, © 2007, with permission from Elsevier: 223 left

Reprinted from *Precambrian Research*, vol. 342, issue 105650, Yin, L. et al, "Microfossils from the Paleoproterozoic Hutuo Group, Shanxi, North China," p.643, fig.6, © 2020, with permission from Elsevier: 230

Reprinted by permission from Springer Nature Customer Service Centre GmbH: *Nature*, Tetrapod trackways from the early Middle Devonian period of Poland, G. Nied wiedzki et al, © 2010: 217

Republished with permission of John Wiley and Sons, from Philippe Janvier, "The Sensory Line System and Its Innervation in the Osteostraci (Agnatha, Cephalaspidomorphi)," *Zoologica Scripta*, vol. 3, 91–99, 1974, fig. 3, permission conveyed through Copyright Clearance Center, Inc.: 171

Riga, A, et al, "In situ observations on the dentition and oral cavity of the Neanderthal skeleton from Altamura (Italy)," *PLOS ONE*, 2020, 15(12): 203

Robert Anemone, courtesy of: 225

Ron Blakey ©2016 Colorado Plateau Geosystems Inc.: 240

Royal College of Physicians, London (via BHL): 58

The Royal Society, London: 152, 153, 166, 174

RR Auction, Boston MA, courtesy of: 96

Ryan Somma, via Flickr (CC BY-SA 2.0): 155

Science Museum, Trento, Matteo De Stefano/MUSE: 84

Science Photo Library, London: 177 (Élisabeth Daynes/S. Entressangle); 106 (E.R. Degginger); 163 (Javier Trueba/MSF); 162 (John Reader); 54–55, 52 top, 70, 124 both (Natural History Museum, London); 69 (Natural History Museum, London/Dr Paul Taylor); 94–95 (Paul D. Stewart); 175 (Roger Harris); 92, 119 (Sinclair Stammers); 170 bottom (Stephen J. Krasemann); 168 (Yale Peabody Museum of Natural History/Martin Shields)

Smithsonian American Art Museum, Washington D.C.; Museum purchase from the Charles Isaacs Collection made possible in part by the Luisita L. and Franz H. Denghausen Endowment: 63 top

Smithsonian Libraries, Washington, courtesy of (via BHL): 1, 3, 14, 17, 27, 32, 33, 34, 36–37, 41, 51, 56, 68, 78–79, 82, 101, 103, 116–117, 118, 135, 150

Smithsonian Institution, Department of Paleobiology: 112 (USNM V2580), 120 (92-6989)

Smithsonian, National Museum of Natural History, Paleobiology Dept, Washington D.C.: CC0: 121, 199 top

SNSB—Staatlichen Naturwissenschaftlichen Sammlungen Bayerns, with kind permission: 242 bottom

State Museum of Natural History, Stuttgart, © SMNS: 189 (photo: Rainer Schoch); 210 (photo: Uli Stuebler)

Stephanie Pierce, courtesy of, after S. E. Pierce, J. A. Clack, and J. R.Hutchinson, "Three-dimensional limb joint mobility in the early tetrapod Ichthyostega." *Nature 486*, no. 7404 (2012): 523–526. https://www.nature.com/articles/nature11124: 146–147

Studio 252MYA / 252mya.com/ Fabrizio di Rossi: 209

Teylers Museum Haarlem, The Netherlands: 28

Thewissen, J. G. M., "From Land to Water: the Origin of Whales, Dolphins, and Porpoises." *Evolution: Education and Outreach*, 2009, no. 2 issue 280. DOI:10.1007/s12052-009-0135-2, after: 216

Timothy Rowe, courtesy of; after T.B. Rowe et al, "X-Ray computed tomography datasets for forensic analysis of vertebrate fossils," *Nature Scientific Data*, 2016, vol. 3, issue 160040, figs 5, 7a: 206, 207

Timothy Rowe and Matt Colbert, University of Texas at Austin, 2010, courtesy of: 223 right

Tom Parker, via Wikimedia Commons (CC by 4.0): 242 top

© Topfoto/Sputnik Images: 214

University of California Libraries (Archive.org, via BHL): 22, 57, 88, 91

University of Connecticut Libraries (Archive.org, via BHL): 8, 9, 148

University of Kansas Biodiversity Institute & Natural History Museum, Lawrence: 182–183, 190

University of Maryland, Baltimore Digital Archive: 11

University of Michigan (via BHL): 60, 143

University of Texas at Austin, Open access (CC BY 3.0): 29

University of Toronto (via BHL): 10 (St Michael's College, John M. Kelly Library); 74 (Gerstein Science Information Centre); 111 (Mississauga); 126–127, 140, 144, 145 (York University)

Vincent Santucci, Courtesy of; after V. L. Santucci et al, "Crossing Boundaries in Park Management," *Proceedings...11th Conference on Research and Resource Management in the Parks & Public Lands, George Wright Society Biennial Meeting*, 2001: 187

Visible Earth/NASA; map adapted from Anemone et al., 2011, by Robert Simmon: 224

Wellcome Collection, London (CC BY 4.0): 2, 7, 15, 38, 39, 75

Western Australian Museum; photo: courtesy of Professor John Long: 198

W.W. Norton & Company, Inc. All rights reserved: 185

Yale University Library, Images of Yale individuals (RU 684), Manuscripts and Archives: 98

Yamagata, K.et al. "Signs of biological activities of 28,000-year-old mammoth nuclei in mouse oocytes visualized by live-cell imaging," *Science Reports* 9, 4050 (2019), https://doi.org/10.1038/s41598-019-40546-1 Open Access (CC BY 4.0): 215

Yousuke Kaifu, photograph courtesy of: 211

**Quadrupedally launching *Pteranodon*. Illustration by James Brown.**

# INDEX

Page numbers in *italics* refer to illustrations

Natica Michelini
Terebratula obovata
T. digona
Ostrea Meadii
(young)
T. flabellum
Parkinsoni
Trochotoma obtusa
Terebratula laginalis
T. maxillata
T. globata
Rhynchonella
Herveyi
Trochus ornatus
T. perovalis
T. fimbria
Rhynchonella spinosa
Purpuroidea Morrisii
T. bullata
T. coarctata
R. concinna
Humphriesianus
Plesiosaurus dolichodeirus
Hybodus reticulatus
(Tooth)
(Tooth)
(Tooth)
Hybodus plicatalis
Saurichthys
Posidonia minuta
Voltzia heterophylla
Dapedius politus